LIFE'S LITTLE RHUBARB COOKBOOK:

101 Rhubarb Recipes

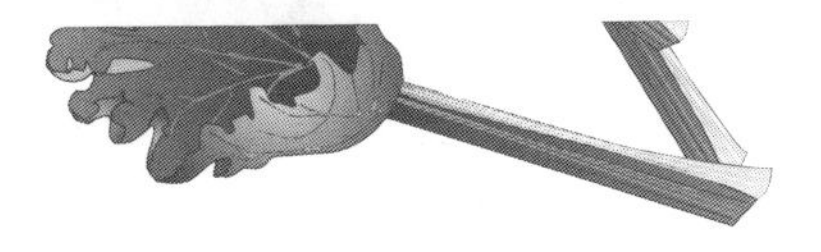

by Joan Bestwick

Life's Little Rhubarb Cookbook
101 Rhubarb Recipes

by Joan Bestwick

ISBN# 1-892384-00-0
Library of Congress Card #99-071836

Published by
Avery Color Studios, Inc.
Gwinn, Michigan 49841

Cover photo by Falcon Photography,
Marquette, Michigan

Proudly printed in U.S.A.

Table of Contents

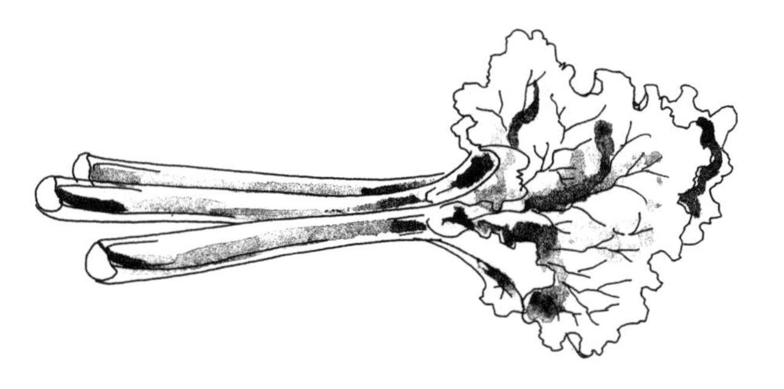

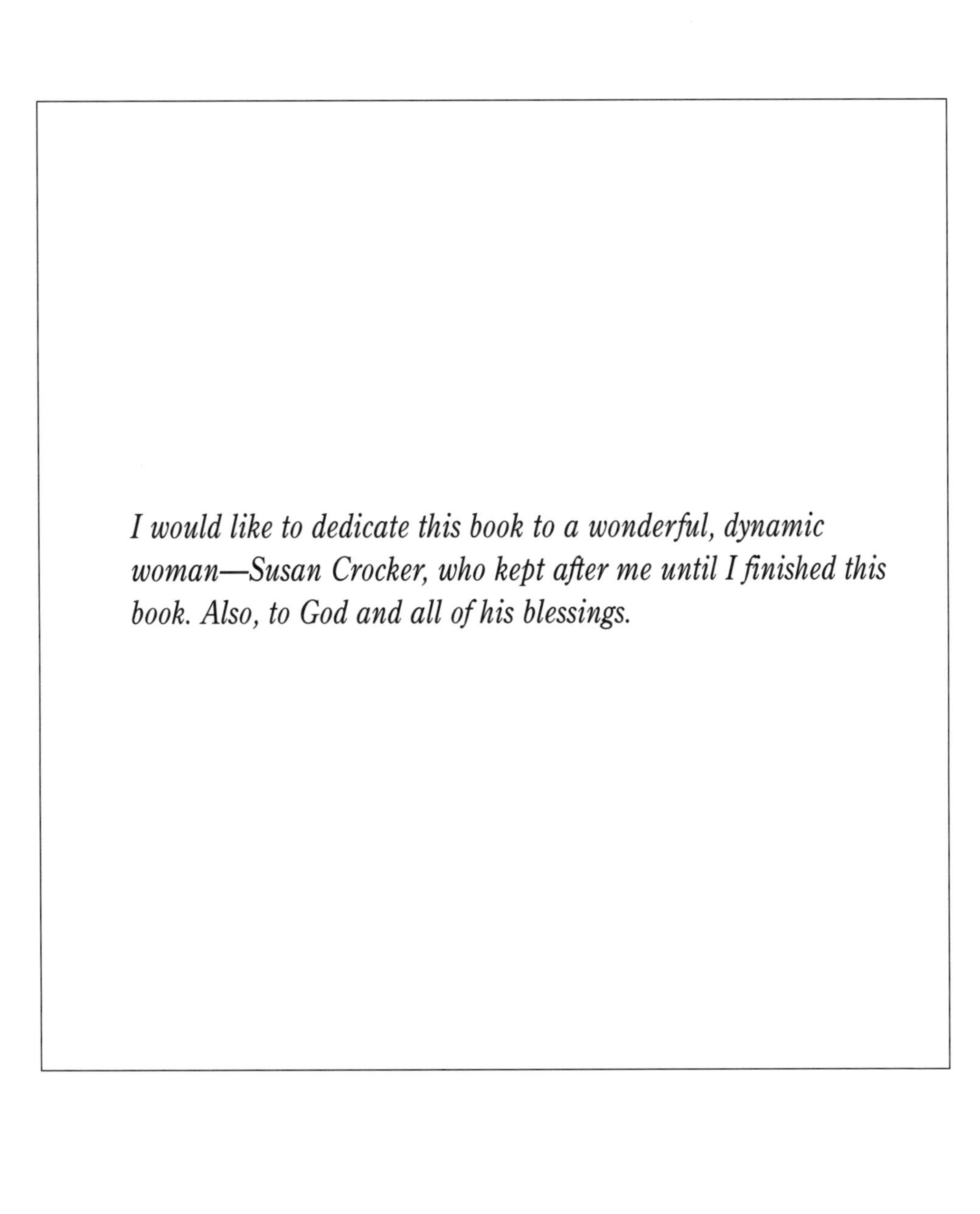

I would like to dedicate this book to a wonderful, dynamic woman—Susan Crocker, who kept after me until I finished this book. Also, to God and all of his blessings.

Always remember to be a creative cook, experiment in the kitchen, play with the recipes and most of all, enjoy God's fruits of the Earth.

Joanie

Rhubarb Facts:

Rhubarb is a root plant that has thick, juicy red stems and large coarse green leaves. It originally grew wild in Asia over 2000 years ago where the Chinese culture used the dried roots as a medicine. A fact that may be unknown to many people is that the leaves are highly poisonous because they contain oxalic acid.

Rhubarb popping out of the ground is a sure sign of spring. To harvest this crop, you need to twist the stalk at the bottom as not to injure the plants root system. Cut the leaf off of the top of the stalk, wash and it is ready to be eaten raw, or it can be frozen, baked, stewed or cooked down. It is possible to force the growth of rhubarb by placing an open ended box around the plant so that the leaves reach for the sunlight, growing taller and faster.

The sour, tart flavor of rhubarb will be sure to awaken all of your senses as well as pucker your lips and cross your eyes. It's a mouth watering experience. Rhubarb is also known as the "pie plant" to many people, but there are many other tasty recipes that you can create with it. Rhubarb is not just for pies, it can be made into sauces, cobblers, jams, crisps, cakes, conserves, salads, drinks, tarts and puddings. The sauces can be put over meats and poultry to perk up their flavor. The jams are great on bagels with cream cheese or warm croissants. My son likes crackers with peanut butter and rhubarb jam.

Fresh rhubarb raw: 8 oz. = 20 calories
1 cup = 31 calories
Fresh, cooked, sweetened rhubarb, 1 cup = 381 calories
Frozen, cooked, sweetened rhubarb, 1 cup = 386 calories

1 pound of rhubarb is 16 oz. or 3 to 4 cups depending on the type of cut used
1 quart of rhubarb is 4 cups
1 pound of sugar is 2-1/4 cups
1 squeezed lemon makes 2 to 3 tablespoons of lemon juice

Chop: To cut in pieces about the size of a pea with a knife, chopper or blender.

Dice: To cut food in small cubes of uniform size and shape.

Slice: A thin piece cut from a larger object, to cut across an object.

Raw: This is uncooked or fresh.

Fresh: Product just produced or harvested.

Frozen: Preserved by freezing.

Young rhubarb: Small tender thin stalks of the rhubarb plant.

Cooked rhubarb: To prepare the rhubarb by placing in 2 tablespoons of water and applying heat and cooking down the product from the raw form.

Stewed rhubarb: To cook by simmering or boiling in water over low heat until product is tender or broken down.

Sweeten: To add sugar to your product to enhance a sweeter taste.

Pureed rhubarb: To cook rhubarb down in water, then to push the food through a sieve or a food processor to make a sauce.

The Original Rhubarb Recipe

Take raw rhubarb, remove the leaf top, trim off the bottom, eat raw—if desired, dip in salt or sugar. The easiest, simplest rhubarb recipe in the book.

JAMS, PRESERVES & CONSERVES

Our value is determined not by what we have
but by what we do with what we have.

Rhubarb Jam

2 quarts chopped rhubarb cut into 1/2 inch slices
6 cups sugar

Put all ingredients into a heavy kettle and cook until thick. Pour hot into sterile jars and seal.

Citrus Pineapple Rhubarb Jam

1-16 oz. package frozen rhubarb, thawed
1-20 ounce can crushed pineapple
1 teaspoon grated orange peel
1 teaspoon grated lemon peel
2 tablespoons lemon juice
6 cups sugar
1/2 bottle liquid fruit pectin
Few drops red food coloring

In a large, heavy saucepan, combine rhubarb, pineapple, grated orange and lemon peel and lemon juice. Add sugar and mix thoroughly. Bring to full boil over high heat and boil rapidly for 1 minute, stirring constantly. Remove from heat; stir in fruit pectin and food coloring. Skim foam, then stir for another 10 minutes to cool jam slightly and keeping fruit in suspension. Ladle jam into hot sterilized jars and seal.

Raspberry Rhubarb Jam

1 quart rhubarb, chopped into 1/2 inch slices
1 quart raspberries
6 cups sugar

Put all ingredients in a heavy, medium saucepan over medium-high heat. Bring to a boil, stirring until thick. Pour into hot sterilized jars and seal.

Pineapple Rhubarb Jam

1 quart rhubarb, chopped into 1/2 inch slices
1 quart chopped canned pineapple
6 cups sugar

Put all ingredients in a heavy, medium saucepan over medium-high heat. Bring to a boil, stirring until thick. Pour into hot sterilized jars and seal.

Currant Rhubarb Jam

1 quart rhubarb, cut into 1/2 inch slices
1 quart currants
6 cups sugar

Put all ingredients in a heavy, medium saucepan over medium-high heat. Bring to a boil, stirring until thick. Pour into hot sterilized jars and seal.

Blueberry Rhubarb Jam

1 quart rhubarb, cut into 1/2 inch slices
1 quart blueberries
6 cups sugar

Put all ingredients in a heavy, medium saucepan over medium-high heat. Bring to a boil, stirring until thick. Pour into hot sterilized jars and seal.

Rhubarb Squash Jam

3 pounds yellow squash
3 pounds rhubarb, cut in 1/2 inch slices
12 cups sugar
3/4 cup chopped crystallized ginger
3 lemons squeezed

Take squash and peel, cut in half, remove seeds and cut into small cubes. In a large bowl layer squash, rhubarb and sugar, and leave overnight. Pour this mixture into a preserving kettle. Add ginger and lemon juice. Keep the rind of the lemon, place in cheese cloth and tie. Place into kettle. Bring the mixture to a rapid boil for 30 minutes or until the jam reaches a setting point. Remove cheese cloth and ladle into hot, sterilized jars and seal.

Ruby Red Jam

5 cups rhubarb, cut into 1/2 inch slices
3 cups sugar
1-3 oz. package raspberry gelatin
1 package raspberry Kool-Aid

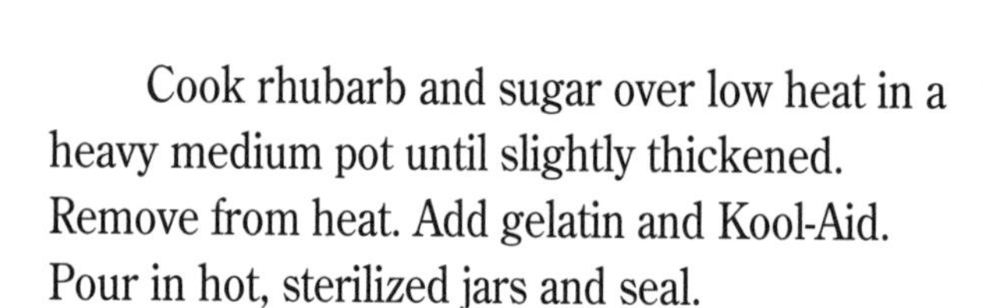

Cook rhubarb and sugar over low heat in a heavy medium pot until slightly thickened. Remove from heat. Add gelatin and Kool-Aid. Pour in hot, sterilized jars and seal.

Jello Rhubarb Jam

5 cups rhubarb, finely cut
4 cups sugar
1-6oz. package strawberry gelatin

In a medium heavy saucepan, mix together rhubarb and sugar. Let it stand overnight. In the morning boil the mixture for 5 minutes. Add gelatin and cook for 3 minutes. Poor into jars and seal.

Variations: Add 1 small can of drained crushed pineapple and 4 cups rhubarb. Cherry, raspberry or lemon-lime gelatin can also be used.

Rhubarb Fig Preserves

2 pounds rhubarb, chopped
1/2 pound figs, chopped
2 lemons, juice and rind
2-1/2 pounds sugar

In a heavy pan or preserving kettle add rhubarb and figs. In separate bowl grate the lemon rind, set aside. In another small bowl squeeze all the juice from the lemon and add the juice to the sugar. Stir this mixture until juice is formed. Place juice in kettle with rhubarb and figs. Cook gently for 30 minutes. Add the grated lemon peel, boil 15 minutes longer until thick and clear. Place in hot, sterilized jars and seal.

Orange Rhubarb Preserves

2-1/2 pounds rhubarb, cut into small pieces
1-1/2 pounds sugar
2 to 2-1/2 tablespoons grated orange peel
3/4 cup orange juice

In a heavy pan combine rhubarb, sugar, orange peel and juice. Stir over low heat until sugar is dissolved. Bring to boil over medium heat. Reduce heat and cook slowly until mixture thickens, about 30 minutes. Stirring occasionally. Ladle into hot sterilized jars and seal. Makes 3 pints.

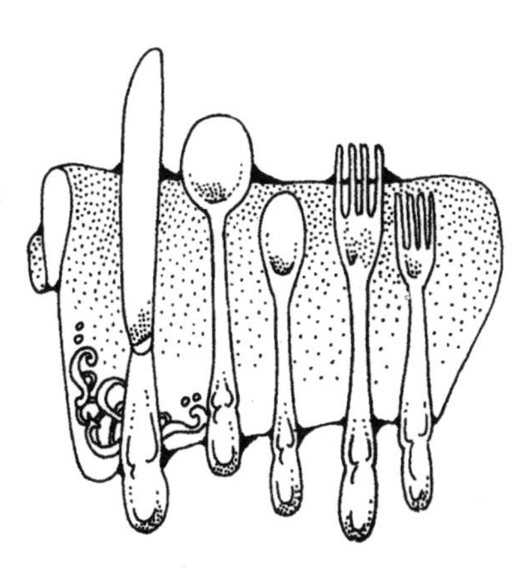

Ginger Rhubarb Conserve

2-16 oz. packages frozen sweetened rhubarb, thawed
1 large orange
1 cup water
1/2 cup golden raisins
3 tablespoons ginger brew
2 tablespoons white vinegar
5-1/2 cups sugar
1/4 teaspoon salt
1 bottle liquid fruit pectin
1/2 cup chopped pecans

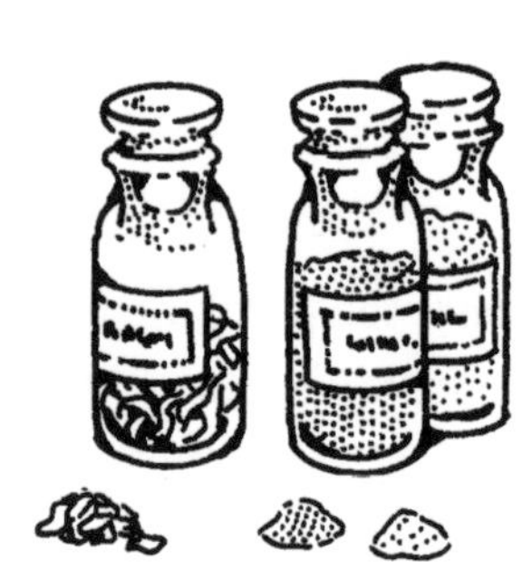

Place rhubarb in a large kettle and set aside. Wash and halve the orange, remove seeds and slice into fine slivers about 3/4 inch long. In a saucepan mix water and orange and simmer about 5 minutes or until peel is almost tender. Add this to the rhubarb. Stir in raisins, ginger brew, vinegar, sugar and salt until blended. Bring to a rapid boil for 1 minute, stirring constantly. Remove from the heat and immediately stir in pectin. Skim off any foam. Add pecans, continue stirring for 5 minutes to keep fruit and nuts in suspension. Ladle into hot sterilized jars and seal.

Ginger brew: Combine 2 teaspoons crushed ginger root with 1/2 cup water in a small saucepan. Cover and bring to boil. Simmer over low heat for 2 minutes. Remove from heat, let stand for five minutes and strain.

Strawberry Rhubarb Conserve

1 quart strawberries, chopped
1 quart rhubarb
7 cups sugar

Clean berries and cut unpeeled rhubarb into 1/2 inch slices before measuring. Place berries and rhubarb into a medium, heavy saucepan, add sugar. Cook slowly until sugar dissolves, then boil rapidly until thick. Pour boiling hot into jars and seal.

Almond Rhubarb Conserve

In a medium heavy pan add
3 cups sliced rhubarb
3 cups white sugar
grated rind and juice of 3 oranges & 1 lemon
1/2 pound chopped almonds

Cook on medium high for 30 minutes. Add 1/2 pound chopped almonds. Cook 5 minutes longer. Place in hot jars and seal.

Rhubarb Pineapple Conserve

6 cups rhubarb
7 cups sugar
2 cups pineapple

Cut unpeeled rhubarb into 1/4 inch slices and cut pineapple into small pieces before measuring. If fresh pineapple is used, cook until tender in just enough water to prevent pineapple from sticking. Combine rhubarb and pineapple with sugar and boil rapidly until thick. Pour boiling mixture into hot jars and seal at once.

SAUCES & RELISH

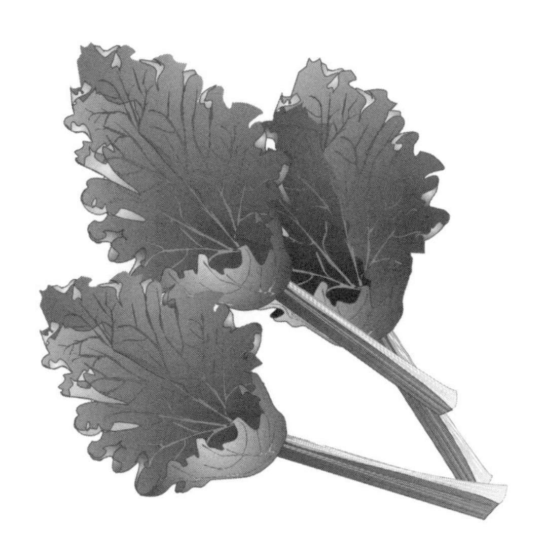

We all have something to teach,
and we all have something to offer.

Rhubarb Sauce

3 cups rhubarb
1 cup sugar
2-3 tablespoons water

Cut rhubarb into 1 inch pieces. In a 2 quart saucepan add rhubarb, sugar, and water. Cover and cook slowly until tender, stirring occasionally.

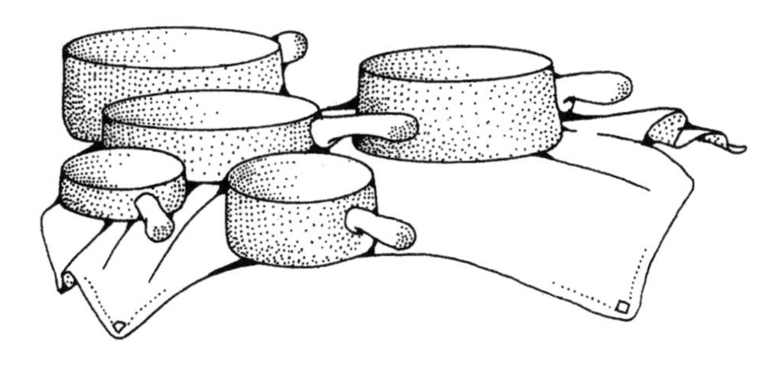

Strawberry Rhubarb Sauce

2-1/2 cups chopped rhubarb
2 tablespoons water
1 cup sugar
2 tablespoons grated lemon peel
1/4 teaspoon salt
1 cup sliced strawberries
2 tablespoons lemon juice
1/4 teaspoon ground cinnamon

In a saucepan over medium high heat combine rhubarb, water, sugar, lemon peel and salt. Bring to a boil, reduce heat. Cook uncovered until rhubarb is soft, about 10-15 minutes. Remove from heat and let stand for five minutes. Stir in strawberries, lemon juice and cinnamon. Cool and serve over cake.

Strawberry Rhubarb Filling

1/2 cup butter or margarine
2 cups fresh sliced strawberries
3 cups fresh sliced rhubarb
2 cups sugar
5 tablespoons cornstarch

In a large heavy saucepan over medium heat, melt the butter or margarine. Add the strawberries, rhubarb and sugar and cook 10 minutes or until soft. Dissolve the cornstarch in a little cold water. Stir this into the fruit and cook until thickened. Cool. Great on short cake or pound cake.

Rhubarb Tonic

2 pounds rhubarb, cut into small pieces
3 cups of water
6 tablespoons sugar

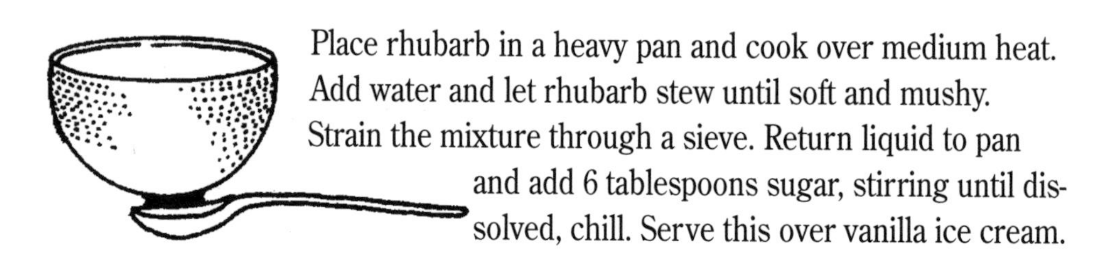

Place rhubarb in a heavy pan and cook over medium heat. Add water and let rhubarb stew until soft and mushy. Strain the mixture through a sieve. Return liquid to pan and add 6 tablespoons sugar, stirring until dissolved, chill. Serve this over vanilla ice cream.

Rhubarb Puree

Cook 3 cups rhubarb in 2 tablespoons water in a heavy, covered saucepan until rhubarb is tender. Force through a sieve and place in a electric blender and blend until smooth.

Raisin Rhubarb Relish

1 cup packed brown sugar
1 cup cider vinegar
1 cup water
1/2 teaspoon allspice
1/2 teaspoon whole cloves
1 stick cinnamon, broken into 4 pieces
1/2 teaspoon celery seeds
1 cup chopped peeled onion
1-1/2 cups sliced rhubarb
1 cup seedless raisins

In an enamel kettle over medium-high heat, combine sugar, vinegar, water, spices and seeds. Boil for 2 minutes, reduce heat to low and add onions and rhubarb. Cover and cook slowly for 30 minutes. Add raisins to hot mixture and cook for 10 more minutes uncovered. Stir a few times with a wooden spoon. Makes 2 pints. Either seal in hot sterile jars or refrigerate.

Rhubarb Chutney

1/2 cup packed brown sugar
1/3 cup vinegar
1/4 cup water
1/4 teaspoon salt
1/4 teaspoon dry mustard
1/8 teaspoon ground cinnamon
Dash of ground cloves
1/2 cup chopped onion
1/4 cup raisins
2 cups chopped rhubarb
3/4 cup pitted whole dates snipped
1/4 cup toasted slivered almonds

In a large saucepan over medium heat, combine brown sugar, vinegar, water, salt, dry mustard, cinnamon and cloves, mixing well. Bring to a boil and stir until sugar dissolves, gently boil uncovered for 5 minutes. Stir in onion and raisins and return to a boil, reduce heat and simmer covered for 20 minutes. Add rhubarb and dates and simmer uncovered for 10 minutes or until thick, stirring occasionally. Add almonds and cool, cover and store in refrigerator for up to 4 weeks.

Rhubarb Chutney II

3 cups fresh chopped rhubarb
2 cups finely chopped onion
2 cups packed brown sugar
2 cups vinegar
1 tablespoon salt
1 teaspoon ground allspice

In a heavy, medium saucepan, stir together all ingredients and bring to a boil. Reduce heat and simmer covered for 30 to 40 minutes or until thick, stirring occasionally. This recipe can also be processed by placing in sterile hot jars, hot water bath for 10 minutes.

PUNCH

Life is a balance between holding on and letting go.

Rhubarb and Lemon Punch

1 small lemon
6 cups fresh or frozen rhubarb, chopped
4 cups of water
2 cups unsweetened pineapple juice
1 small package low calorie lemonade soft drink mix
Ice cubes

Remove the lemon rind from the lemon. Cut and peel into thin slices. In a saucepan combine rhubarb, water and lemon peel. Bring this mixture to a boil, reduce heat and simmer, uncovered for 10 minutes. Carefully strain rhubarb mixture, 1/3 at a time, through a sieve, pressing with a spoon to remove juices. Discard the pulp. Stir pineapple juice and lemonade mix into the rhubarb liquid. Cover and chill. Serve over ice.

Pink Rhubarb Punch

6 cups fresh or frozen rhubarb, 1/2 inch slices
3 cups water
1 cup sugar
1-6 oz. can frozen pink lemonade concentrate
1/4 cup lemon juice
1-1 liter bottle of lemon-lime pop, chilled

In a large saucepan, combine the rhubarb and water. Bring to a boil and reduce heat. Cover and simmer for 5 minutes. Remove from heat. Cool slightly, strain rhubarb mixture pressing to remove all juices. Discard pulp. Add sugar, lemonade, lemon juice to rhubarb juice, stirring to dissolve sugar. Cover and chill.

To serve combine rhubarb juice and lemon-lime pop in a large punch bowl with crushed ice.

Rhubarb Punch

8 cups diced uncooked rhubarb
5 cups water
2 cups honey or sugar
juice from 6 oranges, strained
juice from 3 lemons, strained
1 quart ginger ale chilled

Simmer rhubarb in water until mush. Strain and measure the liquid. Replace liquid back into pan. Add 1/3 cup sugar or honey for each cup of rhubarb juice. Place over low heat, stirring until sugar or honey is dissolved. Cool. Add strained orange and lemon juice. Chill. Before serving add ginger ale. Serve over ice cubes.

SIDE DISHES, SOUPS & SALADS

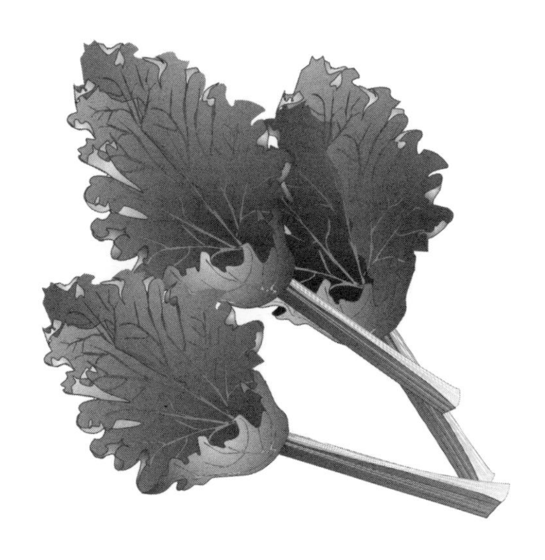

Never underestimate the benefits of being a good listener.

Stewed Rhubarb

Use a 2 quart saucepan with a tight fitting lid.
Place in saucepan:
3 to 4 cups rhubarb, cut into 1 inch pieces

Combine and mix the following:
3/4 cup sugar
1 teaspoon grated lemon peel
1/2 teaspoon cinnamon
2 teaspoons lemon juice

Place over low heat, stirring until sugar dissolves, forming a syrup. Cover and cook slowly about 15 minutes or until rhubarb is tender. You can add a few drops of red food coloring if desired. Serve hot or cold.

Cinnamon Baked Rhubarb

4 cups rhubarb
1-1/2 cups sugar
2 cinnamon sticks
6 cloves
1 orange, sliced

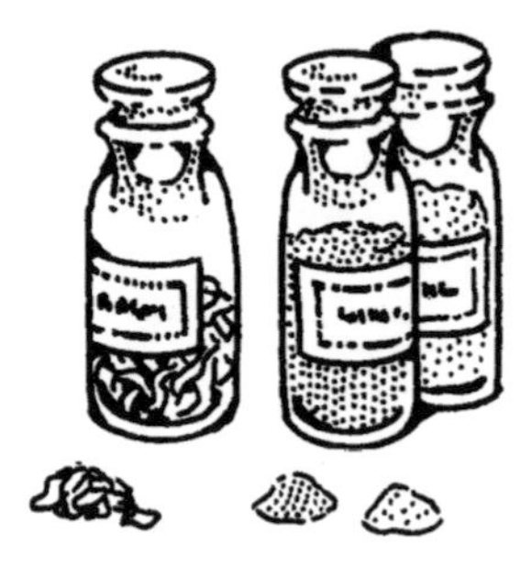

Preheat oven to 350°. In a buttered baking dish put a layer of diced rhubarb. Cover with 1/2 of the sugar, then half of the unpeeled orange slices, one stick of cinnamon and 3 cloves. Repeat layers until the ingredients are used. Place in oven and bake 45 to 50 minutes or until tender.

Scalloped Rhubarb

2 tablespoons butter
2 cups soft bread crumbs
1 cup sugar
juice and rind of 1 orange
1/4 teaspoon cinnamon
1/4 teaspoon nutmeg
3 cups rhubarb, sliced
1/4 cup water

Preheat oven to 375°. Melt butter. In a medium bowl, mix butter and crumbs, then mix sugar, spices and orange rind. Put 1/4 of the crumbs in the bottom of a butter baking dish. Cover with half of the rhubarb. Sprinkle with half of the sugar mixture, then add another 1/4 of the crumbs, the remainder of the rhubarb and the rest of the sugar mixture. Add the orange juice and water. Put the rest of the crumbs over the top. Cover, bake 45 minutes, then uncover and continue to bake until brown.

Chicken with Rhubarb Sauce

1 chicken fryer, cut up
1-1/2 cups rhubarb in 1 inch pieces
2 tablespoons sugar
4 tablespoons honey
1 teaspoon salt
1 cup orange juice
1/4 cup lemon juice
1 tablespoon cornstarch dissolved

In medium saucepan cook rhubarb in orange juice until very soft. Strain through sieve. Return juice to pan and add the rest of the ingredients except cornstarch. Cook until honey and sugar are dissolved. Reserve 1/4 cup of the mixture and add the dissolved cornstarch, stirring until thick and glossy.

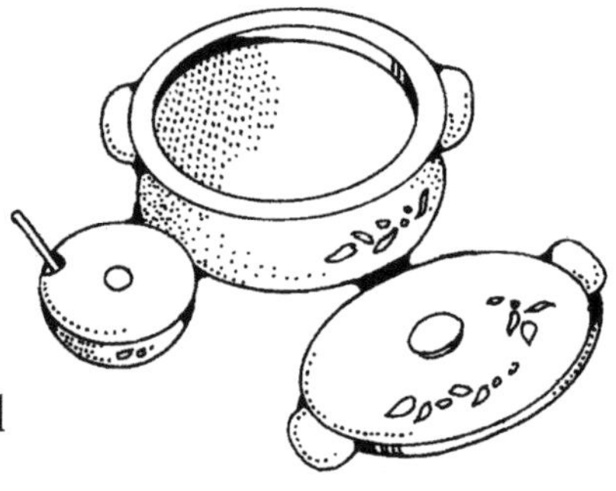

Brown chicken in butter, remove from frying pan and place in casserole dish. Pour thickened sauce over chicken. Bake uncovered at 350° for 40 minutes. Stir reserved rhubarb sauce in casserole and serve.

Stewed Rhubarb and Strawberries

2 pounds rhubarb, cut into 1/2 inch slices
1/2 cup sugar
1 box frozen strawberries, whole

In a heavy saucepan place rhubarb and cover with water. Add sugar when rhubarb is soft, add strawberries. Cook and remove mixture when thickened but strawberries are still whole. Cool and serve.

Baked Rhubarb

1 quart rhubarb, sliced
2 cups of sugar

In a baking dish place rhubarb and sugar. Cover and bake at 350° for 45 minutes, or until rhubarb is tender and deep red in color.

Springtime Rhubarb

3 to 4 cups of sliced 1 inch rhubarb
3/4 to 1 cup sugar
1/2 teaspoon cinnamon
1/2 teaspoon grated lemon peel
2 teaspoons lemon juice

Heat oven to 350°. In a one quart casserole dish toss rhubarb, sugar, cinnamon and lemon peel. Drizzle with lemon juice. Cover and cook for 20 to 25 minutes or until tender. Serve warm or cold.

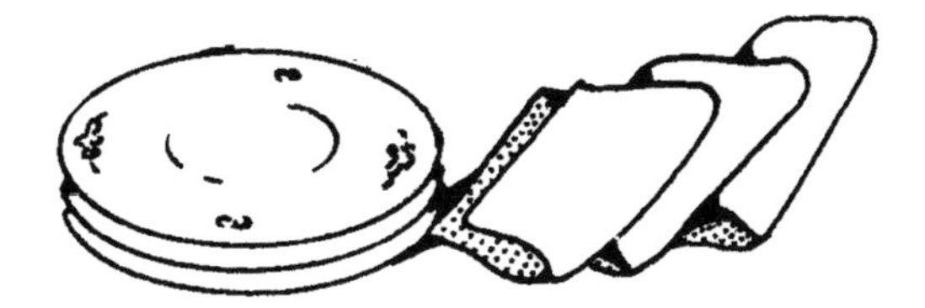

Spiced Rhubarb

3 pounds rhubarb, cut into 1 inch pieces
5 cups sugar
1 teaspoon ground cinnamon
1 teaspoon ground cloves
1 cup vinegar

Place all ingredients into a heavy kettle. Bring to a boil, lower heat and cook slowly until thick. Fill and seal jars

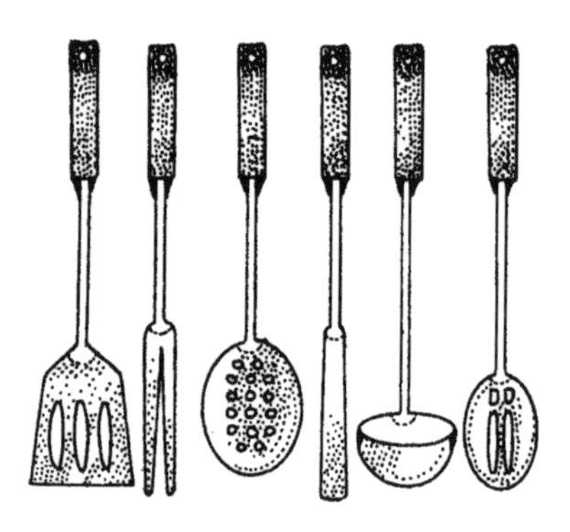

Rhubarb Soup

6 cups water
2 pounds rhubarb cut in 1 inch pieces
1 cup sugar
1 tablespoon cornstarch
1/2 cup cold water

In a soup pot add water and rhubarb, cooking until soft. Add sugar to the rhubarb. In a small bowl mix cornstarch and cold water. Add this to rhubarb and cook until slightly thickened. Serve hot or cold. Makes 8 servings.

Rhubarb with Berries

Take equal parts of rhubarb cut in 1 inch pieces and any fresh berries. Add sugar to sweeten. Let stand one or more hours. Place in saucepan, let mixture slowly heat until sugar is dissolved and cook without stirring until rhubarb is soft. Cool and serve.

Fancy Rhubarb Strawberry Salad

4 cups red rhubarb, cut in 1/2 inch slices
2/3 to 3/4 cup sugar
2/3 cup water
2 boxes (3 oz. each) strawberry gelatin
1/2 cup cold water
1/2 cup white soda or ginger ale
grated rind of 1 orange
1/3 to 1/2 cup chopped walnuts

In a saucepan place rhubarb, sugar, water, and cook until tender. Strain to remove pulp. Add gelatin to mixture, stirring well to dissolve. Add water, soda, and orange rind, chill until partially set. Add nuts and mix. Place in an oiled fancy mold. Refrigerate overnight. Unmold on a lettuce lined glass dish. Serves 12.

Orange Strawberry Rhubarb Salad

4 cups diced raw rhubarb
1-1/2 cups water
1/2 cup sugar
2 packages (3 oz each) strawberry gelatin
1 cup orange juice
1 tablespoon grated orange peel
2 cups fresh sliced strawberries
Strawberries for garnish

Combine rhubarb, water and sugar in saucepan. Cook until tender. Pour hot rhubarb mixture over gelatin and stir until completely dissolved. Add orange juice and peel. Chill until syrupy. Fold in strawberries. Pour into 1 quart mold. Chill until set. Garnish. Serves 8.

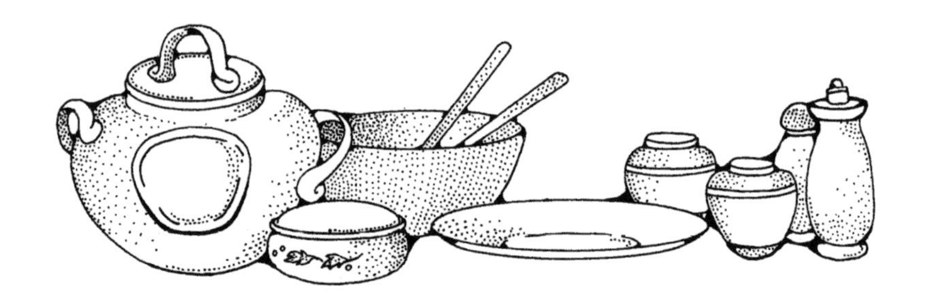

Rhubarb Strawberry Salad

1 large package strawberry gelatin
2 cups hot, cooked rhubarb
2 cups sliced strawberries

To cook rhubarb, place into a medium saucepan with 2 tablespoons water and simmer until tender. Dissolve gelatin in hot cooked rhubarb. Let cool. Then add sliced strawberries. Let set. Serve.

PIES & CAKES

Be patient with the faults of others,
they have to be patient with yours.

Rhubarb Pie

1 recipe for a 9 inch pie crust
1-1/2 to 2 cups of sugar
6 tablespoons flour
4 cups rhubarb, sliced in 1 inch pieces
1-1/2 tablespoons butter or margarine

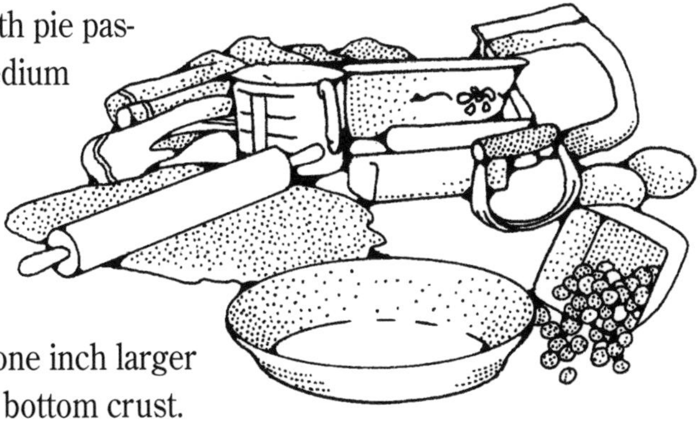

Line 9 inch pie pan with pie pastry. Turn to fit pan. In a medium bowl mix sugar and flour. Combine with rhubarb. Fill pastry lined pan and dot with butter. Cover with top pie crust. Cut several gashes in crust. Trim one inch larger than pan. Fold crust under bottom crust. Press or crimp edges together. Bake in hot oven at 425° for 40 to 50 minutes.

Raspberry Rhubarb Crumb Pie

Filling:
1/2 pint raspberries
1/2 cup honey
5 tablespoons cornstarch
1 teaspoon vanilla
1 teaspoon lemon zest
1-1/4 cups sugar
1/4 cup cold water
7 cups rhubarb, cut into 1 inch pieces
1-9 inch deep dish baked pie shell

Topping:
1 cup coarsely chopped pecans
1/2 cup flour
1/3 cup brown sugar packed
1/3 cup old fashioned oats
1/4 cup cold unsalted butter
1/4 teaspoon ground cinnamon

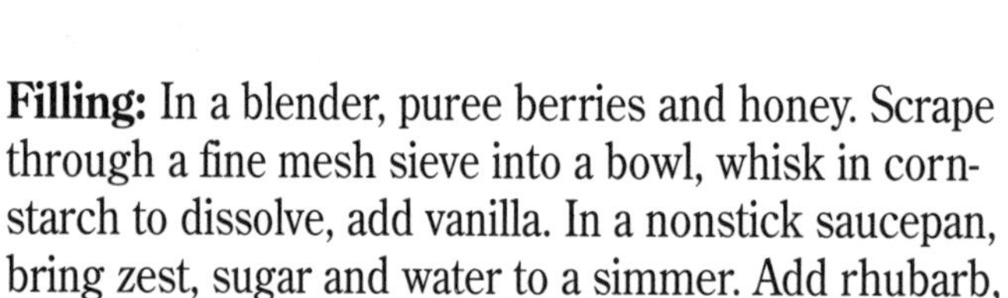

Filling: In a blender, puree berries and honey. Scrape through a fine mesh sieve into a bowl, whisk in cornstarch to dissolve, add vanilla. In a nonstick saucepan, bring zest, sugar and water to a simmer. Add rhubarb, cover and simmer 8-10 minutes, stirring until rhubarb is almost tender. Add berry mixture. Boil 1 minute, stirring until thickened. Pour into pie shell. Preheat oven to 425°.

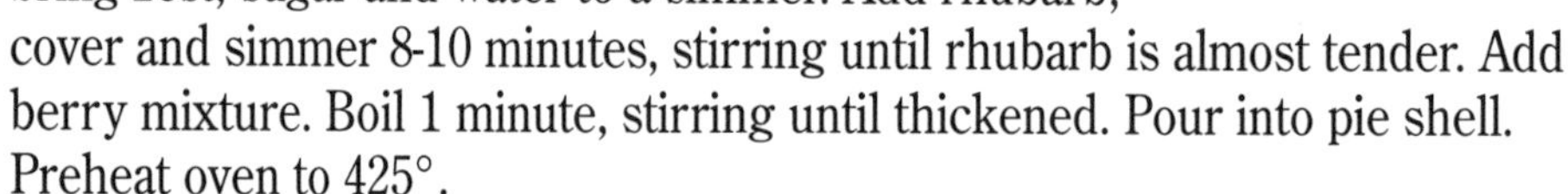

Topping: Combine all ingredients, mix until crumbly. Top pie filling. Place foil on oven rack, then pie. Bake 15 to 20 minutes until browned. Cool on rack.

Rosey Rhubarb Pie

3-1/2 cups rhubarb, chopped into 1/2 inch slices
1 cup strawberry jam
2 tablespoons tapioca
1 teaspoon salt
1-9 inch double pie crust

Preheat oven to 425°. Line a pie plate with bottom pie crust. In a medium bowl, mix rhubarb, jam, tapioca and salt. Pour into pie pan. Top and seal with top crust. Cut vents in top and bake 30 to 35 minutes

Carmel Rhubarb Cobbler

8 tablespoons butter or margarine
3/4 cup brown sugar
1/2 cup white sugar
3 tablespoons cornstarch
1-1/4 cups water
6 cups chopped rhubarb
1-1/4 cups flour
1-1/2 teaspoons baking powder
1/4 teaspoon salt
1/3 cup milk
Cinnamon sugar

In a saucepan over medium heat melt 4 tablespoons of butter or margarine. Add brown sugar, 1/4 cup of the white sugar and cornstarch. To this add the rhubarb and water. Stir and cook until thickened about 5 to 8 minutes. Pour this mixture into a greased 2 quart baking dish and set aside. In another bowl, combine flour, baking powder, salt and remaining sugar. Melt remaining butter, add to dry ingredients with milk. Mix well. Drop by tablespoonfuls onto rhubarb mixture. Bake at 350° for 35 to 40 minutes or until the fruit is bubbly and top is golden brown. Sprinkle with cinnamon sugar.

Strawberry Rhubarb Pie

1 double 9 inch pie crust
3 cups rhubarb, cut into 3/4 inch pieces
2 cups strawberries, sliced 1/2 inch
2 cups sugar
1/3 cup flour
dash of nutmeg
3 eggs beaten
6 tablespoons butter
sugar

Preheat oven to 350°. Place bottom crust into a 9 inch pie pan. In a large bowl combine the rhubarb and strawberries. Place on top the sugar, flour and nutmeg. In the center of the mixture make a well and add the eggs. Use a fork to mix the eggs into the rhubarb and strawberries. Place this mixture into the pie pan and dot with butter. Cover the pie with the top crust, slit the crust and sprinkle with sugar. Bake one hour.

Almost Rhubarb Pie

2-9 inch prepared pie crusts
3/4 cup sugar
1/3 cup flour
1/8 teaspoon salt
1/2 teaspoon vanilla extract
3-1/2 cups rhubarb cut into 1/4 inch pieces
1/4 stick of butter
2 eggs
1/2 cup milk

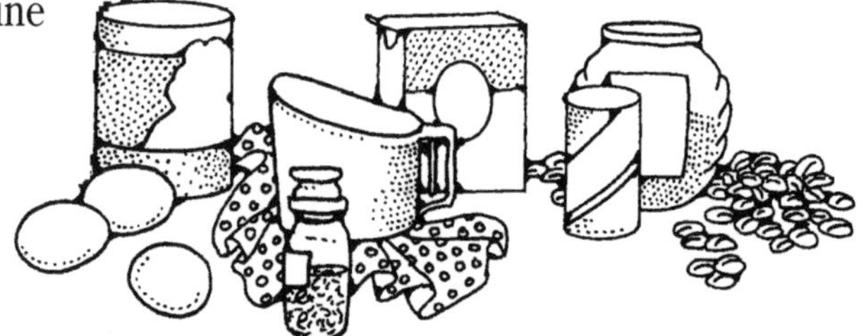

Preheat oven to 450°. Line a 9 inch pie pan with dough. Combine sugar, flour, salt, vanilla and mix well. Add rhubarb, mix and place the filling into the pie shell. Combine 2 eggs and milk. Beat well in a bowl and pour over rhubarb. Dot with butter. Place pie crust on top and cut vents in the crust. Bake pie for 15 minutes then reduce oven to 350° and continue baking for 30 to 40 minutes.

Rhubarb Pancake Cobbler

Base:

1 cup sugar
1/3 cup pancake mix
9 cups rhubarb, cut in 1/2 inch pieces

Combine sugar, pancake mix and rhubarb, toss lightly. Place in a 13 x 9 x 2 inch pan.

Topping:

3/4 cup pancake mix
2/3 cup sugar
1 egg beaten
1/4 cup melted butter or margarine

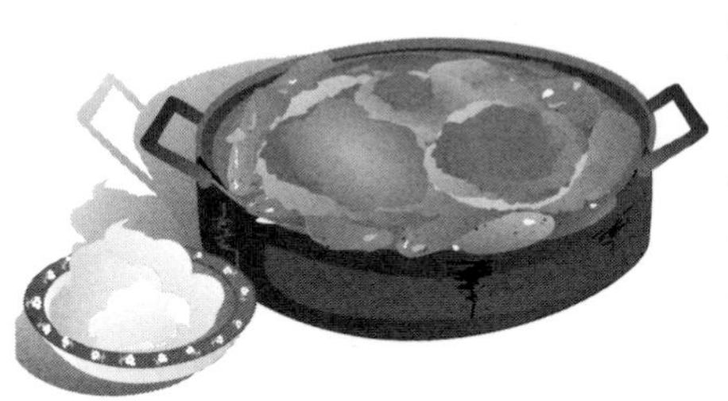

In a bowl combine pancake mix and sugar. Stir in egg until mixture resembles coarse crumbs. Sprinkle evenly over rhubarb base. Then drizzle with melted butter. Bake at 375° for 35 to 40 minutes.

Rhubarb Bavarian Cream Pie

2 cups sliced rhubarb
1/2 cup plus 2 tablespoons honey, divided
1/4 cup water
1 tablespoon unflavored gelatin softened in 2 tablespoons cold water
3/4 cup heavy cream
Red food coloring (optional)
1-9 inch pie crust baked

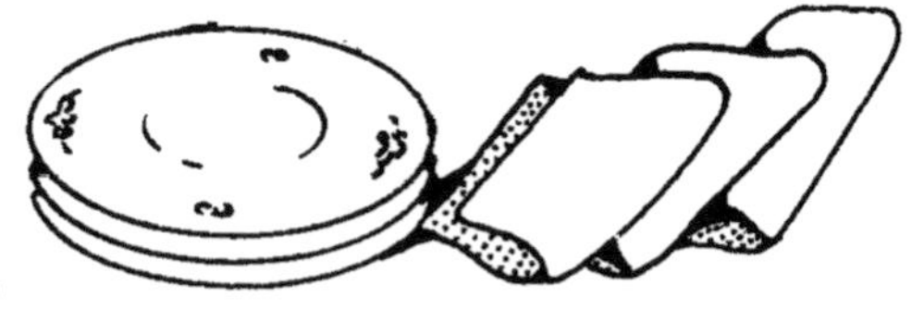

In a medium saucepan, combine rhubarb, 1/2 cup honey and water. Cook until rhubarb is soft. Measure while hot, if necessary, add hot water to make 2 cups fruit mixture. Stir softened gelatin mixture into hot rhubarb. Chill until partly set, but not firm. In a large mixing bowl, whip cream until stiff peaks form. Gradually blend in remaining 2 tablespoons honey. Fold in partly set rhubarb mixture. Add a few drops of red food coloring if desired. Pour filling into pie crust and chill until firm.

Rhubarb Apple Pie

6-8 large apples, cored and sliced
5 cups young tender rhubarb
2 cups sugar
2 tablespoons flour
1-1/2 teaspoons nutmeg
1-1/2 teaspoons cinnamon
pastry for 2-9 inch pies
enough pie dough for a lattice crust

Preheat oven to 350°. Cook rhubarb in a little water until soft and drain well. Mix all ingredients in a large bowl. Divide into 2-9 inch unbaked pastry shells. Top with a lattice crust and bake at 350° for 40 to 50 minutes.

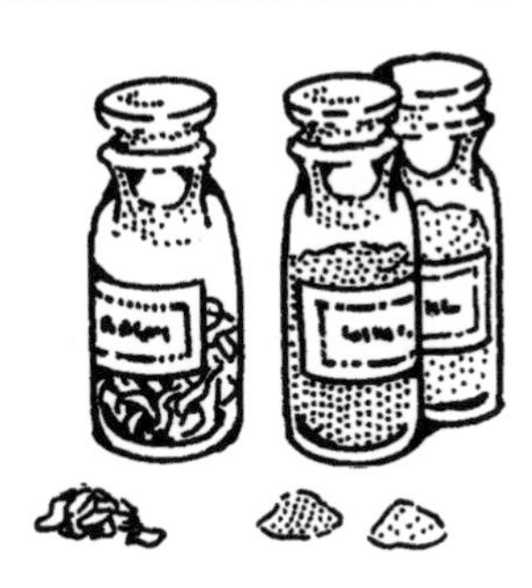

Rhubarb Cream Delight

Crust:
1-1/2 cups flour
3 tablespoons sugar
3/4 cup butter or margarine

Meringue:
4 egg whites
1/4 cup sugar

Cream Filling:
2 cups sugar
4 egg yolks
2/3 cup cream or evaporated milk
3 tablespoons flour
1/2 teaspoon nutmeg
4 cups chopped rhubarb

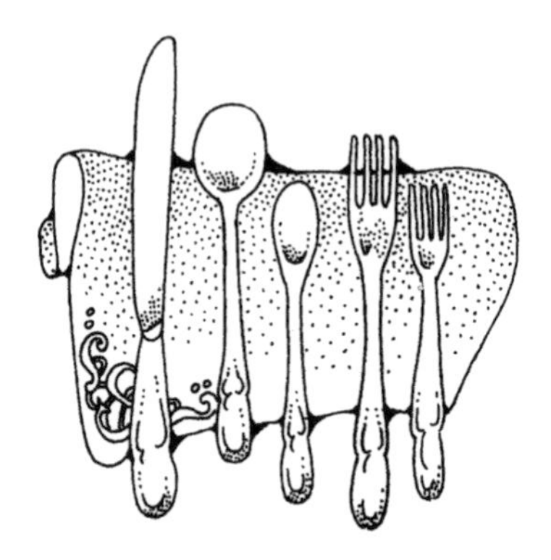

Combine crust ingredients until crumbly; press into a 13 x 9 x 2 inch baking pan. Bake at 350° for 20 minutes. While crust is baking, combine all filling ingredients and cook in heavy saucepan over medium heat. Stir constantly until thickened. Pour hot filling into crust, top with meringue made by beating egg whites with sugar until thick and looks smooth. Bake at 325° for 15 to 20 minutes or until golden brown. Refrigerate leftovers. Makes 10-12 servings.

Rhubarb Strawberry Orange Pie

1 unbaked deep dish pie shell
enough pie dough for a lattice crust

Filling:
3 cups sliced fresh rhubarb (cut into 1/4 inch pieces)
3 cups sliced fresh strawberries
1/2 to 3/4 cup sugar
1-1/2 tablespoons instant tapioca
1/3 cup fresh orange juice
1-1/2 tablespoons orange marmalade (optional)
1/4 teaspoon orange peel

Preheat oven to 400 °. Combine filling ingredients in large mixing bowl, let stand for 15 minutes while tapioca softens. Pour filling into pie shell. Prepare lattice strips for top crust. Bake at 400° for 20 minutes, reduce heat to 375° and bake 30 minutes until rhubarb is tender. Makes 6-8 servings.

Rhubarb Strawberry Pie

2 cups rhubarb sliced into 1 inch pieces
2 cups sliced strawberries
1-1/2 cups sugar
Pinch of salt
1/3 cup flour
1/2 teaspoon almond extract
2 tablespoons butter
1 prepared 9 inch pie crust

In a medium bowl, mix together gently, strawberries, flour, rhubarb, sugar, salt and almond extract. Put into pastry shell. Dot with butter. Cover with pastry crust and brush top crust with milk and sprinkle with sugar. Bake at 400° for 40 to 50 minutes until fruit is tender and crust is lightly brown.

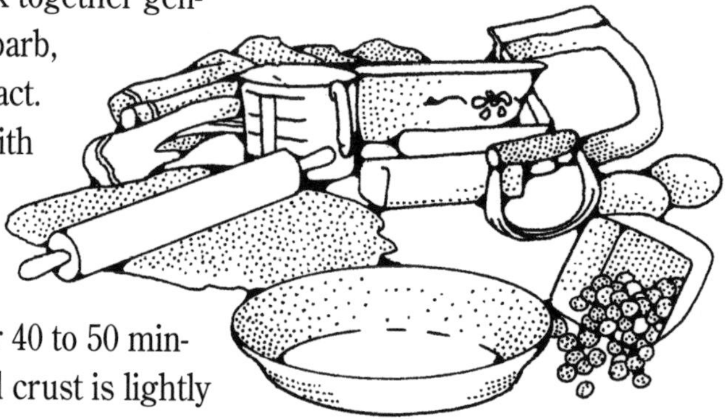

Rhubarb Orange Cream Pie

1-9 inch baked pie shell

Filling:
1-1/2 cups sugar
2 tablespoons cornstarch
3 cups fresh rhubarb, cut in 1/2 inch pieces
1/2 cup cream, half & half or milk
1/4 cup orange juice
5 drops red food coloring
3 egg yolks, slightly beaten

Meringue:
3 egg whites
6 tablespoons cream of tartar
1/4 teaspoon sugar
1/2 teaspoon vanilla

In a medium saucepan, combine sugar, cornstarch, rhubarb, cream, orange juice and food coloring. Cook over medium heat stirring frequently until rhubarb is tender and mixture has thickened. In another saucepan, pour 1 cup hot rhubarb mixture into egg yolks, stirring constantly. Add the rest of the hot rhubarb mixture, bring to boil. Cool slightly, pour filling into pie shell. Make meringue by beating egg whites and cream of tarter until soft peaks form. Slowly add sugar and vanilla beating until stiff peaks form. Spread over filling sealing edges. Bake at 350° for 12 minutes or until golden brown.

Rhubarb Sour Cream Custard Pie

3 cups rhubarb, cut in 1/4 inch pieces
3 tablespoons flour well rounded
1 cup sugar
1-9 inch unbaked pie shell
3 eggs separated
1 tablespoon thick dairy sour cream

Topping:
1-1/2 cups old fashioned quick cooking oats
1 cup brown sugar
1 teaspoon cinnamon
1/4 cup butter or margarine

Place cut rhubarb in a large mixing bowl. Combine flour and sugar in a separate bowl and add to rhubarb, mix and let stand while preparing the crust. Place pie dough in pie plate and flute the edges high in order to hold the filling. Brush the crust with egg white wash from separated eggs. Beat egg yolks and sour cream until thick, add to rhubarb mixture, pour into pie shell. Combine topping ingredients, spread evenly over pie. Bake at 400° for 10 minutes. Reduce heat to 350° and bake for 50 minutes more.

Rhubarb Crunch

1 cup whole wheat flour
1 cup brown sugar
1 teaspoon cinnamon
3/4 cup oatmeal
1/2 cup melted butter
4 cups diced rhubarb

In a medium bowl mix all ingredients except the rhubarb. When crumbly, place 1/2 of the mixture into a 9 inch baking pan and cover with the rhubarb.

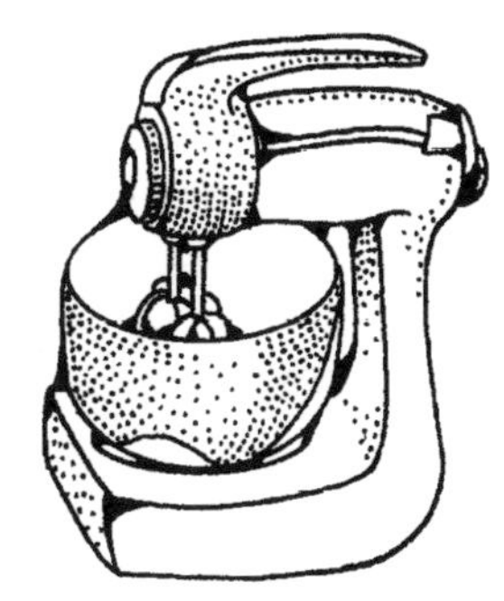

In a small pan combine:
1 cup sugar
2 tablespoons cornstarch
1 cup water
1 teaspoon vanilla

Cook this mixture on medium-low until clear. Pour this over the rhubarb. Top it with the remaining crumb mixture and bake at 350° for 45 minutes or until the rhubarb is tender.

Rhubarb Cream Pie

1-1/2 cups sugar
3 tablespoons flour
1/2 teaspoon nutmeg
1 tablespoon butter
2 well beaten eggs
3 cups sliced rhubarb
1-9 inch pie pastry

In a medium bowl blend sugar, flour, nutmeg, butter and eggs. In a pastry lined 9 inch pie pan place rhubarb. Place top of pastry over rhubarb and seal. Cut slits on top and bake in hot oven at 450° for 10 minutes, then reduce heat to 350° and bake for about 30 minutes.

Rhubarb Custard Pie with Variations

Make pastry for 2-9 inch pie crusts
In a medium bowl beat:
3 eggs slightly
1 tablespoon milk

Mix and stir in:
2 cups sugar
1/4 cup flour
3/4 teaspoon nutmeg
4 cups sliced rhubarb

Pour into pastry lined pie pan. Dot with 1 tablespoon butter. Cover with lattice top pie crust. Bake at 400° for 50 to 60 minutes or until browned.

Variations: #1 use 2 cups rhubarb - 2 cups blueberry, #2: 3 cups rhubarb - 1 cup drained crushed pineapple.

Rhubarb Whipped Cream Pie

2 tablespoons whipped unflavored gelatin
1/2 cup cold water
2-1/2 cups cooked, tender, chopped rhubarb
1 cup sugar
1 cup heavy cream whipped
1-9 inch pie crust, baked

Soften gelatin in water. Heat rhubarb and sugar until boiling in medium pan. Add gelatin and stir until dissolved. Cool. When mixture begins to thicken fold in whipped cream. Pour into baked pie shell and chill. Makes 1-9 inch pie.

Rhubarb Pineapple Pie

3 tablespoons quick cooking tapioca
1-1/4 cups sugar
1/2 teaspoon salt
2-1/2 cups fresh rhubarb, cut into 1/2 inch pieces
1-1/4 cups drained crushed pineapple
1/3 cup water
Pastry for 1-9 inch pie

In a medium bowl, combine quick cooking tapioca, sugar, salt, rhubarb, pineapple and water. Let stand about 15 minutes. Line a 9 inch pie pan with pastry. Trim pastry 1 inch larger that pan. Fold edge to form standing rim and flute. Fill with fruit mixture and sprinkle with crunchy top. Bake at 425° for 35 to 40 minutes or until syrup boils with heavy bubbles that do not burst.

Crunchy Top:
1/3 cup brown sugar
3 tablespoons flour
1/2 teaspoon cinnamon
2 tablespoons softened butter

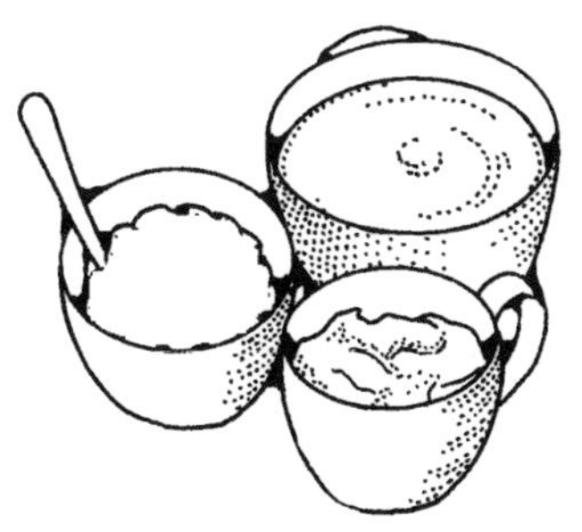

Combine above four ingredients. Mix with pastry blender or fork until crumbs are the size of large peas.

Honey Rhubarb Custard Pie

3 cups cut rhubarb
2 eggs
2 tablespoons milk
1 cup honey
4 tablespoons flour
1/4 teaspoon salt
1/4 teaspoon nutmeg
1-9 inch pie shell
2 tablespoons butter

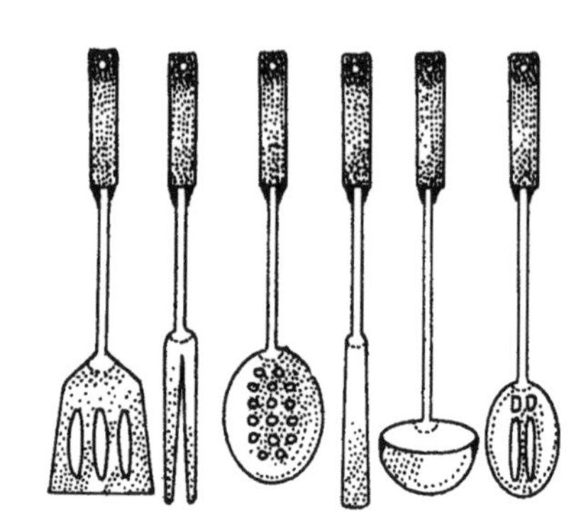

Combine above ingredients in a mixing bowl and mix well. Pour in unbaked 9 inch pie shell. Dot with butter and crumb topping.

Crumb Topping:

1/3 cup brown sugar
3 tablespoons flour
1/2 teaspoon cinnamon
2 tablespoons butter

Combine above ingredients, mix with pastry blender until crumbs are the size of large peas. Bake at 400° for 50 to 60 minutes.

Rhubarb Cobbler

4 cups rhubarb
1 tablespoon butter
1 tablespoon quick tapioca
1 tablespoon grated lemon rind
1 teaspoon ground cinnamon
2 cups flour
2 teaspoons double acting baking powder
1/2 teaspoon salt
1 tablespoon sugar
4 tablespoons butter
3/4 cup half and half or milk

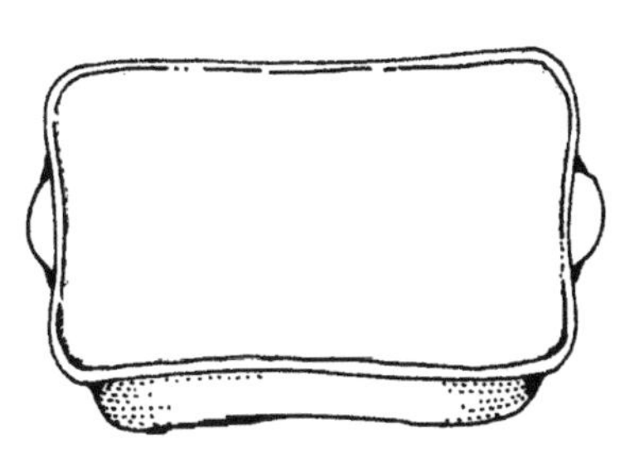

Place rhubarb in a medium saucepan and add 2 tablespoons of water, cook on medium-low heat until tender, sweeten with sugar to your taste. Put rhubarb in buttered cake pan. Dot butter over rhubarb and sprinkle in tapioca. Sprinkle cinnamon and lemon rind over top. In a bowl put flour, baking powder, salt and sugar. Mix. Then with a fork add butter and mix. Add gradually the milk to make dough that holds together but is soft. Drop by spoonfuls onto rhubarb. Sprinkle with additional sugar. Bake at 400° for 30 minutes.

Ma's Rhubarb Treat

3-4 cups rhubarb sliced into 1 inch pieces
1/2 cup white sugar
3 eggs
1 cup heavy cream
1/2 cup flour
1 prepared pie crust

Preheat oven to 375°. Arrange rhubarb pieces in bottom of prepared crust. In a medium bowl beat sugar and eggs until fluffy. Add heavy cream and flour and beat for 3 minutes. Pour this mixture over rhubarb. Bake for 45 minutes or until well cooked and slightly brown. Cool.

Rhubarb Peachy Pie

1 can (8-1/2 oz.) sliced peaches
2 cups chopped rhubarb
1 cup sugar
1/4 cup flaked coconut
3 tablespoons quick-cooking tapioca
1 teaspoon vanilla extract
Pastry for double crust 9 inch pie
1 tablespoon butter or margarine

Drain peaches, reserving syrup and dice. In a bowl place peaches, peach syrup, rhubarb, sugar, coconut, tapioca and vanilla. Line a 9 inch pie plate with bottom pastry. Add filling; dot with butter. Top with remaining pie pastry and flute the edges. Slit top of dough. Bake at 350° for 1 hour or until crust is golden brown and filling bubbles.

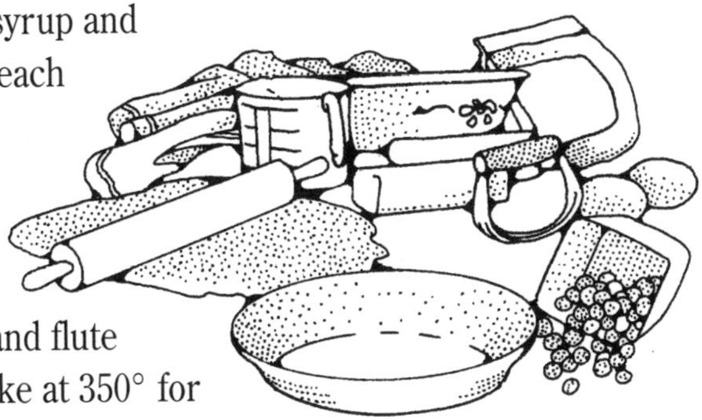

Sour Cream Rhubarb Pie

Pastry for a single pie crust
2 eggs
1 cup sour cream
1-1/2 cups sugar
2 tablespoons flour
1 teaspoon vanilla
1/4 teaspoon salt
3 cups chopped rhubarb
1/4 cup packed brown sugar
1/4 cup flour
3 tablespoons butter or margarine

Roll out the pastry and line a 9 inch pie plate with pastry and flute the edges. Set aside. For filling; in a large mixing bowl, stir together eggs and sour cream. Stir in sugar, 2 tablespoons flour, vanilla, and salt. Stir in chopped rhubarb. Pour the filling into the prepared pastry. Cover the pie with foil. Bake at 450° for 15 minutes. Reduce the oven to 350° and bake for 20 minutes. In a small mixing bowl, stir together the brown sugar and 1/4 cup flour. Cut in butter until mixture resembles coarse crumbs. Sprinkle mixture over rhubarb filling.

Bake mixture, uncovered for 20 to 25 minutes until filling is set.

Rhubarb Cake

Base:

2 cups flour
1/2 cup butter
1/2 teaspoon salt
1 teaspoon baking powder
1 beaten egg

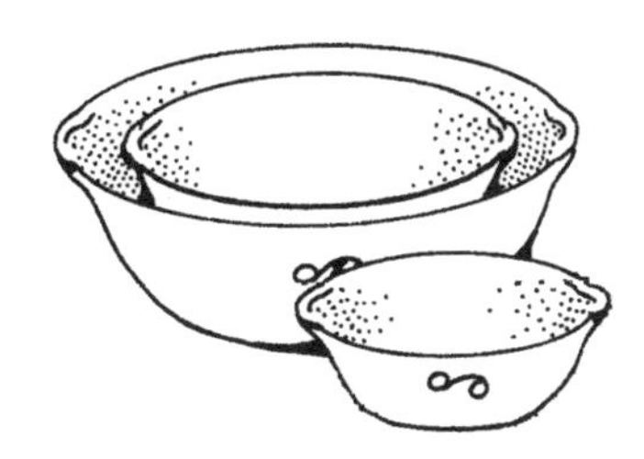

Mix in a medium bowl all ingredients except egg. When mixture is mixed well add beaten egg. Reserve 1 cup of the mixture for topping. Take remaining mixture and flatten into a 9 x 12 inch pan.

Filling:

1-1/2 cups sugar
1/2 cup flour
2 eggs
4 cups chopped rhubarb
1/2 cup melted butter

In a large bowl mix all ingredients in order. Spread on base.

Topping:

Reserved topping
1 cup brown sugar
3 tablespoons cinnamon sugar

Mix well in small bowl. Sprinkle over cake filling. Bake at 350° for 45 minutes to 1 hour.

Buttermilk Rhubarb Cake

1/2 cup softened butter
1-1/2 cups sugar
1 egg
1 cup buttermilk
2 cups flour (reserve 2 tablespoons to mix with rhubarb)
1 teaspoon baking soda
1/2 teaspoon salt
1 teaspoon cinnamon
1 teaspoon vanilla
2 cups diced rhubarb fresh or frozen

In a medium bowl cream together butter, sugar and egg. Add buttermilk alternately with flour, soda, salt, cinnamon. Mix until smooth. Add vanilla. Toss rhubarb with remaining flour and add to mixture. Spread in greased and floured 13 x 9 x 2 inch cake pan. Bake at 350° for 35 minutes.

Rhubarb Upside Down Cake

4 cups rhubarb, cut into 1 inch pieces
1 cup sugar
1 cup miniature marshmallows
1-3/4 cups flour
2 teaspoons baking powder
1/8 teaspoon salt
1/2 cup vegetable shortening
1 cup sugar
1/3 teaspoon almond extract
1/3 teaspoon vanilla extract
2 eggs separated
1/2 cup milk

Cook rhubarb, covered in a heavy saucepan over low heat until juice begins to run. Remove from heat. Add one cup sugar and marshmallows, stir until marshmallows are melted. Pour in greased 9x 9 cake pan. Sift together flour, baking powder and salt, and set aside. Cream shortening with one cup sugar and extracts. Add egg yolks and beat vigorously. Beat in dry ingredients and milk alternately. Beat egg whites until stiff and fold into mixture. Pour over rhubarb. Bake at 350° for 40 to 50 minutes. Loosen cake from sides of pan and turn onto cake plate immediately.

Rhubarb Pudding Cake

Cake:

1 cup sugar
1 egg
2 tablespoons butter or margarine melted
1 cup buttermilk or sour milk
1/2 teaspoon salt
1/2 teaspoon baking soda
1 teaspoon baking powder
2 cups all purpose flour
1 cup diced fresh rhubarb

Topping:

2 tablespoons margarine, melted
1/2 cup sugar

Vanilla Sauce:

1/2 to 1 cup sugar
1/2 cup margarine
1/2 cup evaporated milk
1 teaspoon vanilla extract

In a medium bowl blend together sugar, egg and butter. Beat in buttermilk until smooth. Stir together in separate bowl the dry ingredients, mix well. Add this to the buttermilk mixture. Mix well and stir in rhubarb. Pour this into a 9 inch square pan. In a small bowl combine topping ingredients, sprinkle on top of batter. Bake at 350° for 45 minutes or until cake tests done. For sauce, mix together sugar, margarine and milk in a saucepan. Boil one minute, stirring constantly. Remove from heat, stir in vanilla. Serve sauce over cake. 12 servings.

Rosey Red Rhubarb Cake

1/4 cup butter
2 cups flour
2-1/2 teaspoons baking powder
1/4 teaspoon salt
1/4 cup brown sugar packed
1 egg, slightly beaten
3/4 cup milk
6 cups thinly sliced rhubarb
1 package (3 oz) strawberry gelatin

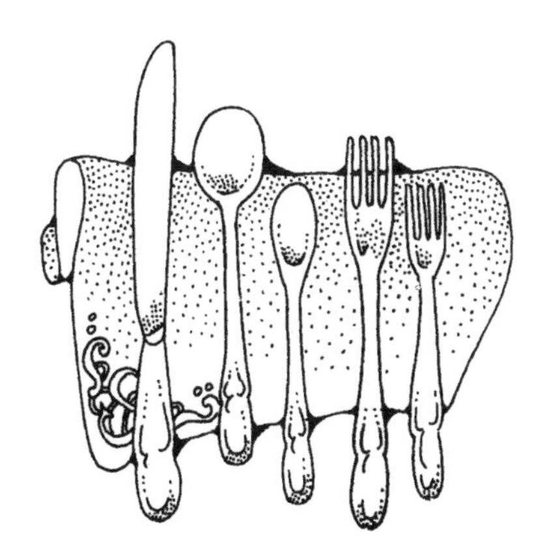

Topping:
6 tablespoons butter
1-1/2 cups sugar
1/2 cup flour

In a medium bowl cut butter into flour, baking powder, salt and brown sugar as for pie crust. Add egg and milk, mix well. Spread into a 13 x 9 x 2 inch baking pan. Mixture will be moist. Top with rhubarb and sprinkle powdered gelatin mix over rhubarb. Combine topping ingredients like pie crust. Sprinkle over the top of rhubarb. Bake at 350° for 60 minutes. Serve warm or cold. 12 servings.

Rhubarb Almond Pudding Cake

2 cups sliced rhubarb
3/4 cup sugar
4 tablespoons margarine
1 cup flour
1 teaspoon baking soda
1/2 cup milk
1/2 teaspoon almond extract
1 cup boiling water

Place rhubarb in a greased 8 x 8 inch pan. Cream sugar and margarine set aside; mix flour and soda and stir into creamed mixture with milk and almond extract. Spread batter over rhubarb. Pour boiling water over batter. Bake at 325° for 50 minutes.

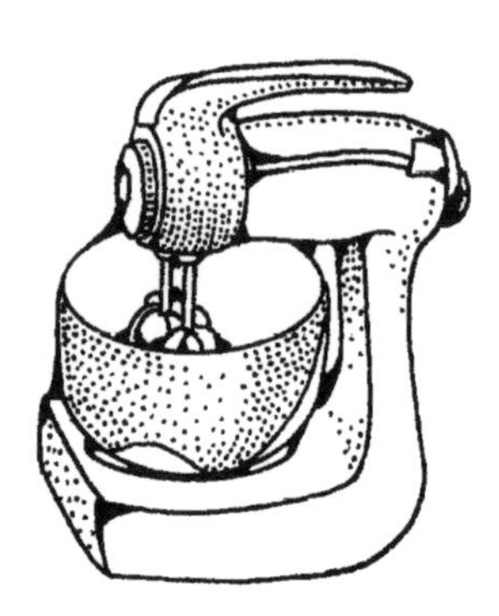

Rhubarb Bottom Cake

3 tablespoons melted butter
2/3 cup sugar
2 cups cut up rhubarb
1/2 cup raisins

Melt butter in an 8 x 8 inch pan. To this add remaining ingredients.

Cake:

3 tablespoons melted butter
1/2 cup sugar
1 egg
1-1/2 cups flour
2 teaspoons baking powder
1/2 teaspoon salt
1/2 cup milk

Mix butter, sugar and egg. Add dry ingredients with milk and mix until smooth. Pour batter over topping in pan. Bake at 375° for 25 minutes. Invert pan. Serve with whipped topping.

Sweet-Tart Shortcake

Baked short cakes
Slivered almonds
2 cups halved strawberries
1-1/2 cups rhubarb
1/4 cup sugar
whipped cream

Place rhubarb in a medium saucepan with 2 tablespoons of water, cook on medium-low until tender and sweeten with sugar to your taste. In serving bowls place split short cakes. Sprinkle with the almonds. In a medium bowl mix strawberries, rhubarb and sugar. Fill shortcakes with this mixture and top with whipped topping.

Gooey Upside-Down Cake

Topping:
3 cups sliced fresh rhubarb
1 cup sugar
3/4 cup flour
2 tablespoons ground nutmeg
1/4 cup melted butter or margarine

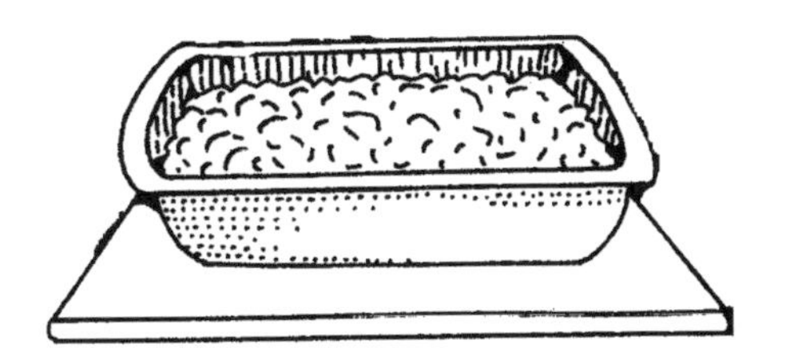

Batter:
1-1/2 cups flour
3/4 cup sugar
2 teaspoons baking powder
1/4 teaspoon salt
1/2 teaspoon ground nutmeg
1/4 cup melted butter or margarine
2/3 cup milk
1 egg
whipping topping

Sprinkle rhubarb in a greased 10 inch heavy skillet. Combine sugar, flour and nutmeg. Sprinkle over rhubarb and drizzle with butter. For batter, combine flour, sugar, baking powder, salt and nutmeg in a mixing bowl. Add butter, milk and egg. Beat until smooth. Spread over rhubarb mixture. Bake at 350° for 35 minutes or until cake tests done. Loosen edges immediately and invert onto serving dish. Serve warm with whipped topping.

Rhubarb Egg Cake

1 pound rhubarb, sliced 1/2 inch
2/3 cup sugar
1/2 cup margarine
1/2 cup flour
1-1/3 cups water
3 tablespoons sugar
1/4 cup blanched chopped almonds
2 eggs separated
1 cup whipped topping

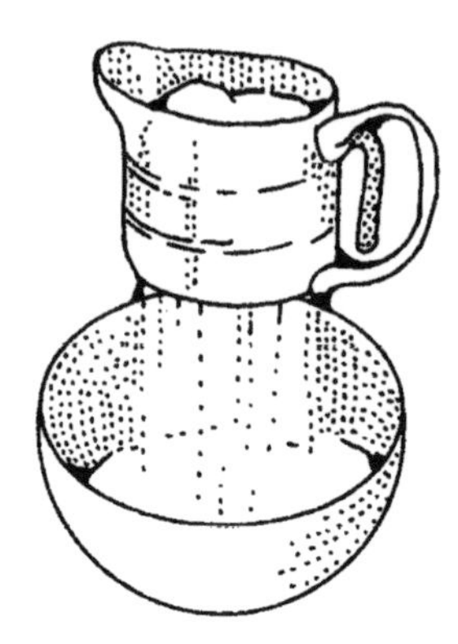

Preheat oven to 400°. In the bottom of a greased casserole or baking pan place rhubarb. Sprinkle 2/3 cup sugar over rhubarb. Melt margarine, add sifted flour and water. Mix and cool. Beat egg yolks and add to flour mixture and beat well. Add the 3 tablespoons sugar and almonds. Fold in stiffly beaten egg whites. Pour batter over rhubarb and bake until rhubarb is soft and crust is brown. Serve with whipped topping.

Rhubarb Yogurt Cake

Topping:

1/2 cup packed brown sugar
1/2 cup chopped pecans
1/4 cup flour
1 cup whole wheat flour
1-1/2 teaspoons baking powder
2 tablespoons butter or margarine
1 teaspoon ground cinnamon

Cake:

1 cup flour
1 cup whole wheat flour
1-1/2 teaspoons baking powder
1 teaspoon ground cinnamon
1/2 teaspoon baking soda
1/2 teaspoon salt
1 - 8 oz. carton cherry vanilla yogurt
3 tablespoons milk
1/2 cup butter or margarine
1 cup packed brown sugar
2 eggs
1 teaspoon vanilla
2-1/2 cups chopped fresh or frozen rhubarb thawed

Preheat oven 350°. Grease and lightly flour a 13 x 9 x 2 inch pan. Set aside. For crumb topping: In a small bowl, stir together 1/2 cup brown sugar, chopped pecans, and 1/4 cup flour, whole wheat flour, baking powder, 1 teaspoon cinnamon. Mix well and set aside.

For cake: In a medium mixing bowl stir together 1 cup flour, whole wheat flour, baking powder, cinnamon, baking soda and salt. Set aside. In a large mixing bowl beat the 1/2 cup butter with mixer on medium speed for 30 seconds. Add brown sugar and beat until fluffy. Add eggs and vanilla and beat well. Add the dry ingredients and yogurt, alternating to the beaten mixture, beating on low speed. Fold in rhubarb. Pour rhubarb mixture into prepared pan. Sprinkle the crumb topping evenly over the batter. Bake for 40 to 45 minutes or until cake test done.

Whole Wheat Rhubarb Pudding Cake

5 cups sliced rhubarb
1/4 cup sugar
1-1/4 cups all purpose flour
1 cup sugar
1/2 cup whole wheat flour
1/2 cup chopped pecans
1-1/4 teaspoons baking powder
1/2 teaspoon ground cinnamon
1/8 teaspoon ground nutmeg
1/4 teaspoon salt
3/4 cup milk
1/4 cup margarine, melted
1-1/4 cups sugar
1 tablespoon cornstarch

Preheat oven to 350°. In a 3 quart rectangular baking pan, mix rhubarb and 1/4 cup sugar. In a bowl mix flour, 1 cup sugar, whole wheat flour, pecans, baking powder, cinnamon, nutmeg and 1/4 teaspoon salt. Stir in milk and margarine. Spread batter over rhubarb. Mix 1-1/4 cup sugar and cornstarch. Add 1-1/4 cup boiling water until sugar dissolves. Slowly pour over batter. Bake for 45 minutes until done. It may be necessary to cover cake the last 10 minutes to prevent over browning. Serve warm.

DESSERTS & PUDDING

"Give a blessing to someone today, the effects will go on like a pebble thrown into water." Joan Bestwick

Rhubarb Mandarin Crisp

6 cups chopped rhubarb
1-1/2 cups sugar
5 tablespoons quick cooking tapioca
1 can (11 oz.) mandarin oranges, drained
1 cup packed brown sugar
1 cup quick cooking oats
1/2 cup flour
1/2 teaspoon salt
1/2 cup butter or margarine

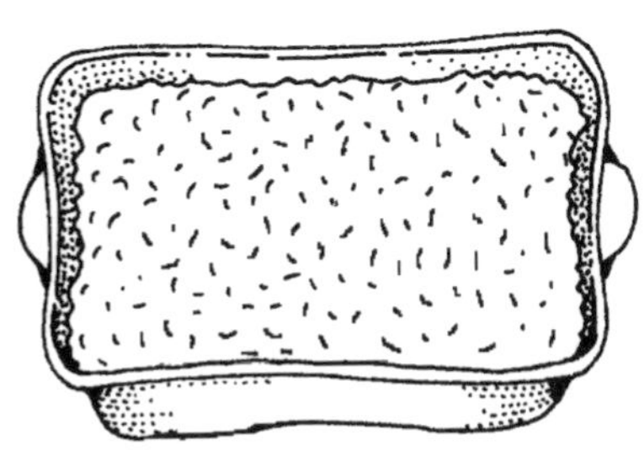

In a medium bowl toss rhubarb, sugar and tapioca. Let stand for 15 minutes, stirring occasionally. Pour into a greased 13 x 9 x 2 inch baking pan. Top with oranges. In a bowl, combine brown sugar, oats, flour and salt. Cut in butter until mixture resembles coarse crumbs. Sprinkle evenly over oranges. Bake at 350° for 40 minutes or until top is golden brown.

Rhubarb Cream Delight Dessert

Crust:
1 cup flour
1/4 cup sugar
1/2 cup butter or margarine

Rhubarb layer:
3 cups fresh rhubarb, cut in
1/2 inch pieces
1/2 cup sugar
1 tablespoon flour

Cream layer:
12 oz. cream cheese
1/2 cup sugar
2 eggs

Topping:
1 cup (8 oz.) sour cream
2 tablespoons sugar
1 teaspoon vanilla extract

For Crust: Mix crust ingredients well in a bowl and pat mixture into 10 inch pie plate. Set aside. For Rhubarb layer: Combine all ingredients. Toss lightly and pour into crust. Bake at 375° for about 15 minutes. While this is baking prepare cream layer by beating cream cheese and sugar until fluffy. Beat in eggs one at a time. Pour this mixture over hot rhubarb layer. Bake at 350° for 30 minutes until set. Combine topping ingredients. Spread over hot layers. Chill. Makes 12-16 servings.

Rhubarb Pizza

1 to 2 cups rhubarb puree
sugar to taste
A double 9 inch pie crust
sliced strawberries (pepperoni)
sliced or chopped pistachio nuts (green olives)
dried cherries (black olives)
grated white chocolate (cheese)

To make rhubarb puree: cook rhubarb on low heat in a medium saucepan until soft and add sugar to taste. Place rhubarb mixture in food processor and puree.

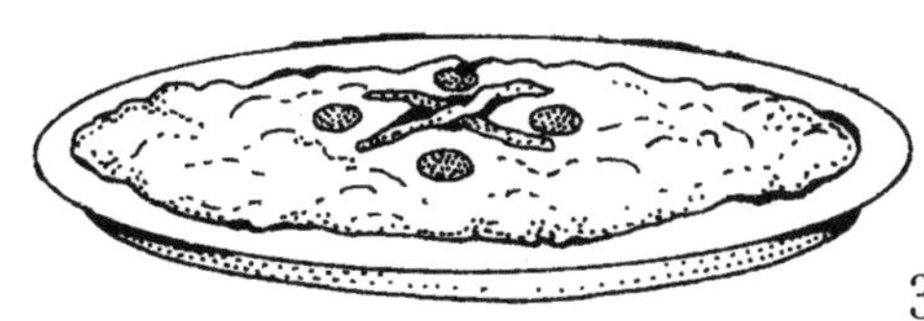

On a cookie sheet place rolled out pie crust. Pour puree on top of crust to cover crust like pizza sauce. Bake at 375° until crust is done. Remove from oven. Take toppings and top the crust. Sliced Strawberries (pepperoni), Sliced or chopped pistachio nuts (green olives), dried cherries (black olives), grated white chocolate (cheese). Tastes great served with vanilla ice cream.

Rhubarb Bread

1-1/2 cups brown sugar
2/3 cup oil
1 egg
1 cup sour milk or buttermilk
1 teaspoon baking soda
1 teaspoon salt
1 teaspoon vanilla
2-1/2 cups flour
1-1/3 cups rhubarb cut fine
1/2 cup chopped nuts

In a large bowl mix all ingredients in order. Pour into greased and floured bread pan.

Topping:
Mix in small bowl
1/2 cup sugar
1 tablespoon butter

Sprinkle on top of bread mixture. Bake at 350° for about 1 hour.

Rhubarb Strawberry Crunch

1 cup flour
1 cup brown sugar packed
3/4 cup quick cooking oats
1 teaspoon ground cinnamon
1/2 cup butter
4 cups fresh or frozen sliced rhubarb
1 pint fresh strawberries, halved
1 cup sugar
2 tablespoons cornstarch
1 cup water
1 teaspoon vanilla extract

Preheat oven to 350°. In a bowl combine flour, brown sugar, oats and cinnamon. Cut in butter until crumbly. Press 1/2 of mixture into a 9 inch square baking pan. Combine rhubarb and strawberries; spoon over crust. In a saucepan, combine sugar and cornstarch. Stir in water and vanilla. Bring to boil over medium heat. Cook and stir for 2 minutes. Pour over fruit. Sprinkle with remaining crumb mixture. Bake for 1 hour. Great with ice cream.

Rhubarb Cinnamon Muffins

1-1/2 cups flour
1/2 cup plus 1 tablespoon sugar
2 teaspoons baking powder
1-1/4 teaspoons ground cinnamon
1/4 teaspoon salt
1 egg beaten
2/3 cup buttermilk
1/4 cup butter or margarine melted
1/2 cup chopped rhubarb
1/4 cup peach preserves

In a medium bowl combine flour, 1/2 cup sugar, baking powder, 1 teaspoon cinnamon and salt. In a small bowl, combine egg, buttermilk and butter. Stir this mixture into the dry ingredients just until moistened. Spoon 1 tablespoon of batter into 9 greased or paper-lined muffin cups. Combine rhubarb and preserves. Place 1 tablespoonful in center of each cup (do not spread). Top with remaining batter. Combine remaining sugar and cinnamon, sprinkle over batter. Bake at 400° for 20 minutes or until top of muffin springs back when lightly touched in the center. Makes 9 muffins.

Rhubarb Fritters

1 cup all-purpose flour
1 cup plus 1 tablespoon sugar
1/2 teaspoon salt
2 eggs separated
1/2 cup milk
1 tablespoon melted butter
2 cups finely chopped rhubarb
oil for deep frying
confectioners sugar

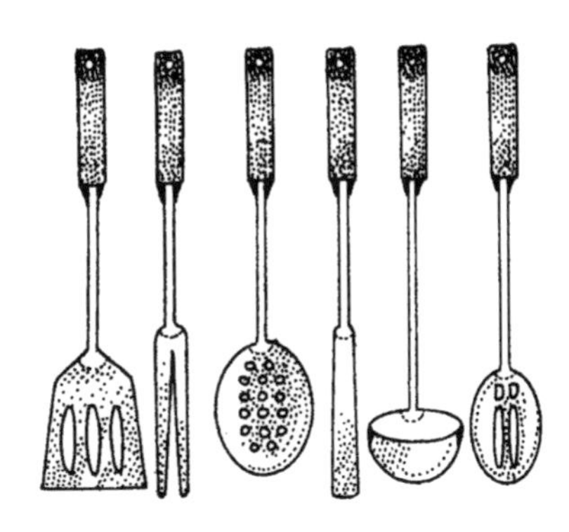

In a medium bowl combine flour, 1 cup sugar and salt. In another bowl, whisk egg yolks, milk and butter. Gradually add this mixture to the dry ingredients, stirring until smooth. Toss in remaining sugar; gently stir into batter. In a mixing bowl, beat egg whites until stiff, fold into batter, add rhubarb and mix well. In an electric skillet or deep fryer, heat oil to 375°. Drop batter by tablespoon into oil. Fry a few at a time, turning with a slotted spoon until golden brown. Drain on paper towels. Dust with confectioners sugar, serve warm.

Apple Rhubarb Crisp

2 cups finely cut apples
2 cups finely cut rhubarb
1 egg beaten
3/4 cup white sugar
1/4 teaspoon nutmeg
1/2 cup butter or margarine
1 cup flour
1 cup brown sugar

Mix together the apples, rhubarb, egg, white sugar and nutmeg in a medium mixing bowl. Place this mixture in a glass 9 x 13 x 2 baking dish. Combine the butter, flour and brown sugar into a crumb mixture. Place this over the rhubarb mixture and press down. Bake at 375° for 30 minutes. Serve it with ice cream or whipped topping.

Rhubarb Strawberry Turnovers

1 pint strawberries, cleaned and chopped
1-1/4 cups fresh or frozen rhubarb, sliced into 1/2 inch pieces
3/4 cup sugar
1 teaspoon fresh lemon juice
2 tablespoons cornstarch
1 package (17-1/4 oz.) puff pastry sheets thawed
confectioners sugar for dusting

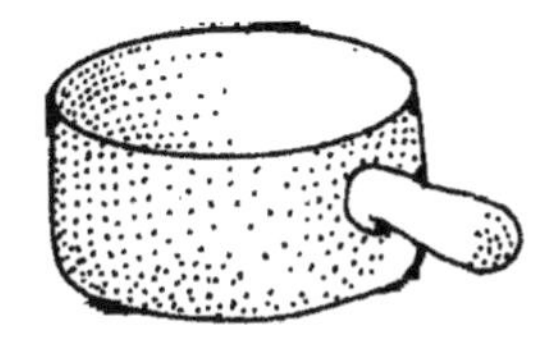

Heat oven to 400°. In a saucepan combine berries, rhubarb, sugar, lemon juice and cornstarch. Cook over medium-high heat for 10 minutes. Reduce heat to medium and cook another 2 minutes or until thickened, clear and bubbly. Let cool slightly. Open pastry sheets on work area. Cut each sheet into 4 equal pieces. Coat 8 muffin tin cups with a non-stick spray. Gently press middle of 1 pastry square into bottom of muffin cup. Fill with almost 1/4 cup of filling. Fold over 4 pastry flaps. One on top of the other, covering the filling. Brush with water to seal. Repeat with remaining puff pastry and filling. Bake at 400° for 15 to 20 minutes, until tops are puffed and golden brown. Cool in pan or on a rack for 10 minutes. Remove from muffin tins. Dust with confectioners sugar.

Strawberry Rhubarb Squares

1 cup packed brown sugar
1/2 cup margarine or butter, softened
1/4 cup shortening
2 cups Bisquick baking mix
1-1/2 cups quick cooking oats
1 package (16 oz.) frozen rhubarb, thawed and well drained or 2-1/2 cups fresh rhubarb cut into 1/2 inch pieces
1 package (10 oz.) frozen sliced strawberries, thawed and well drained or 1 cup fresh sliced strawberries
1 cup sugar
1/4 cup Bisquick baking mix
1 egg
a few drops red food coloring (optional)

Heat oven to 375°. Grease 9x 9x 2 inch pan. Mix brown sugar, margarine & shortening until well blended. Stir in 2 cups Bisquick baking mix and the oats until crumbly. Press half of the crumbly mixture in pan. Bake until set, 10 minutes. Mix rhubarb, strawberries, sugar, 1/4 cup Bisquick baking mix and the egg in a 2 quart saucepan. Heat over medium heat, stirring constantly, until mixture thickens and boils. Boil and stir 30 seconds if using frozen fruit. 2 minutes if fresh fruit. Stir in food coloring. Spread over hot baked layer. Sprinkle remaining crumbly mixture over top, press gently. Bake until golden brown, about 30 minutes. Cut into about 3 inch squares. Serve warm or cool with or without whipped cream or ice cream. Refrigerate remaining squares. Makes 24 squares.

Cranberry Rhubarb Tart

1 can whole berry cranberry sauce
1/3 cup sugar
1-1/2 tablespoons cornstarch
3/4 pound rhubarb, cut into 1/2 inch pieces
Pastry for a 9 inch single pie crust
Powdered sugar

Preheat oven to 375°. Combine cranberry sauce, sugar and cornstarch in a medium mixing bowl. Stir in rhubarb. Pour into a pastry lined 9 inch pie plate. Fold crust edge over the filling, crimping to fit. Bake 40 minutes or until golden brown. Cool completely. Sprinkle with powdered sugar before serving.

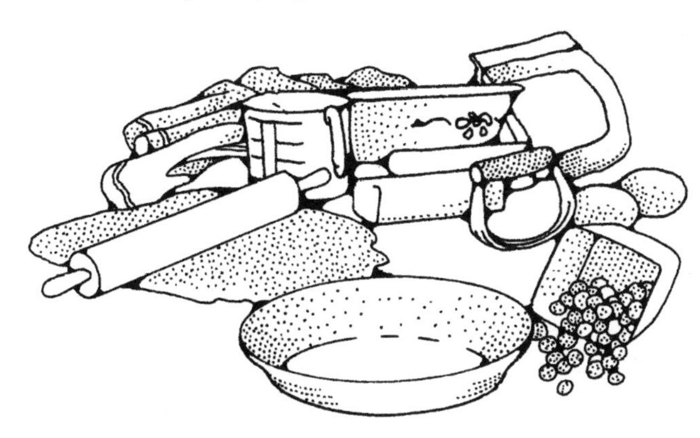

Blueberry Rhubarb Crisp

3 cups fresh or frozen blueberries, (thawed)
2 cups fresh or frozen sliced rhubarb, (thawed)
1/2 cup rolled oats
1/2 cup flour
1/2 cup packed brown sugar
1/2 teaspoon ground cinnamon
1/4 cup butter or margarine

Preheat oven to 350°. In a buttered 2 quart square baking dish place fruit. In a bowl, combine oats, flour, brown sugar and cinnamon. With a fork cut in butter until this mixture resembles crumbs. Sprinkle topping over the fruit. Bake for 30 to 35 minutes or until fruit is tender and topping is golden. Serves 6.

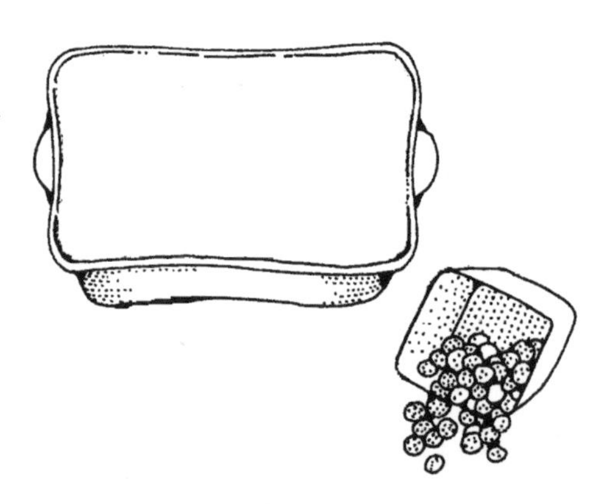

Rhubarb Muffins

1 egg
1-1/4 cups brown packed sugar
1 cup buttermilk
1/2 cup vegetable oil
2 teaspoons vanilla extract
2-1/2 cups flour
1 teaspoon baking soda
1 teaspoon baking powder
1/2 teaspoon salt
1-1/2 cups diced fresh rhubarb
1/2 cup chopped walnuts

Topping
1/3 cup sugar
1 teaspoon ground cinnamon
1 teaspoon butter or margarine, melted

In a mixing bowl, beat egg. Add brown sugar, buttermilk, oil and vanilla. Beat 1 minute. Combine dry ingredients, stir into sugar mixture, just until moistened. Fold in rhubarb and walnuts. Fill greased or paper lined muffin tins 3/4 full. Combine topping ingredients in a small bowl, sprinkle over muffins. Bake at 375° for 20-25 minutes. Makes 1 dozen.

Rhubarb Dream Bars

Crust:
2 cups flour
3/4 cup confectioners sugar
1 cup butter or margarine

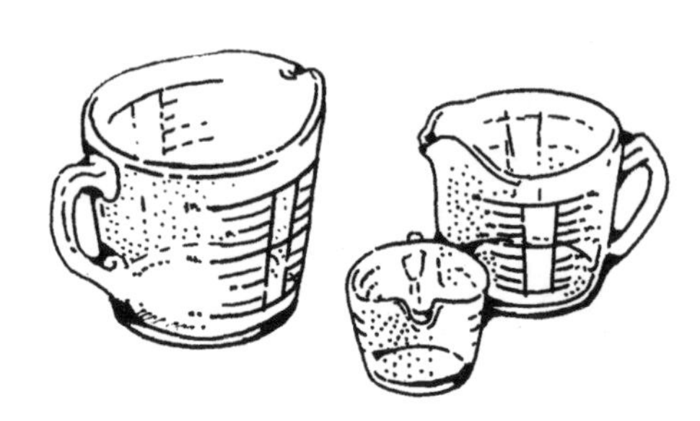

Filling:
4 eggs
2 cups sugar
1/4 cup flour
1/2 teaspoon salt
4 cups thinly sliced, fresh rhubarb

Mix crust ingredients together in a medium bowl. Press into a 15 x 10 inch pan. Bake at 350° for 15 minutes. The crust will be light in color. Combine eggs, sugar, flour and salt. Beat well. Fold in the rhubarb. Spread filling mixture on hot crust. Return to oven to bake 40 to 45 minutes longer. Cool, cut into bars. Yield 36 bars.

Rhubarb Fluffs

7-1/2 cups rhubarb, cut in 1 inch pieces
2/3 cup water
1-1/2 cups sugar
dumpling dough
whipped cream

Dumpling Dough:
2 cups Bisquick
2/3 cup milk

In a 4 quart dutch oven, combine rhubarb, water and sugar, heat to boiling, stirring occasionally. Prepare dumpling dough by mixing ingredients with a fork in a bowl. Drop dough by spoonfuls into boiling rhubarb. Cook uncovered over low heat 10 minutes. Cover and cook 10 minutes longer. Serve warm with whipped cream.

Yummy Rhubarb Crunch

3 cups diced rhubarb
1 cup sugar
3 tablespoons flour

Topping:
1 cup brown sugar
1 cup old-fashioned rolled oats
1-1/2 cups flour
1/2 cup butter
1/2 cup vegetable shortening

In a medium bowl combine rhubarb, sugar and flour. Place in a greased 13 x 9 x 2 inch pan. Combine brown sugar, oats and flour, cut in butter and shortening until crumbly. Sprinkle over rhubarb mixture. Bake at 375° for 40 minutes. Serve warm with ice cream, whipped cream or milk. Makes 10-12 servings.

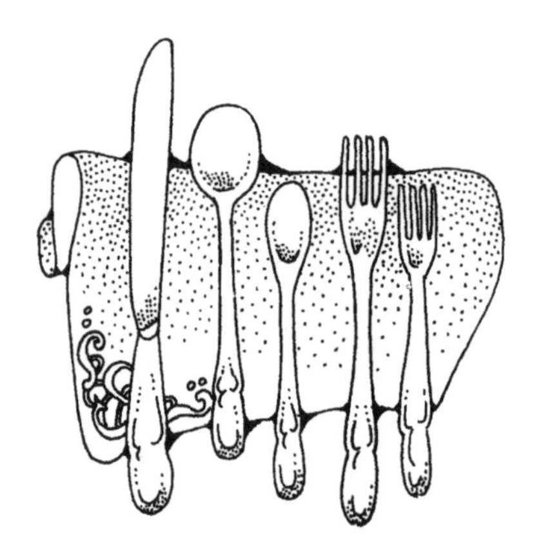

Rhubarb Macaroons

8 macaroons
1 cup thick sweetened rhubarb sauce
whipped cream

Crumble macaroons and divide crumbs into 4 individual serving dishes. Cover macaroons with rhubarb sauce. Top with whipped cream.

Rhubarb Freeze Tarts

4 cups rhubarb, cut in 1 inch slices
3/4 to 1 cup sugar
1/2 teaspoon ground cinnamon
1 teaspoon grated lemon peel
2 teaspoon lemon juice
3 to 4 drops red food coloring
1 pint vanilla ice cream slightly softened
8 tart shells

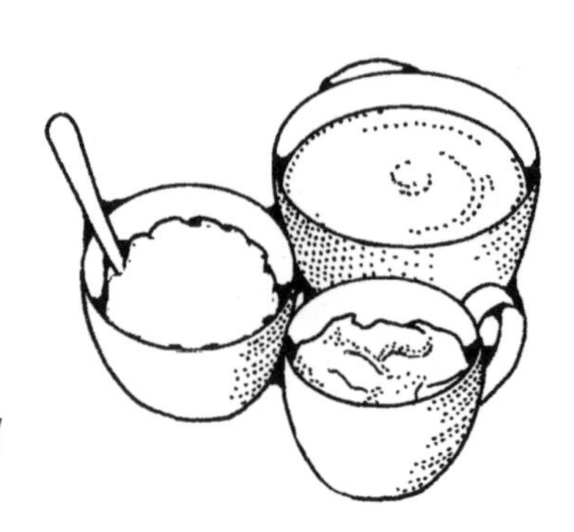

Preheat oven 350°. In a 1 quart shallow baking dish with a cover, toss rhubarb with sugar, cinnamon and lemon peel. Drizzle lemon juice over rhubarb mixture. Bake covered for 20 to 25 minutes. Cool and chill.

When chilled use a blender to chop up rhubarb mixture and tint with food coloring. Gradually blend mixture with the softened ice cream. Turn into molds, cover and freeze. Before serving, set into the refrigerator to soften slightly. Spoon into tart shells and serve.

Rhubarb Compote

2 cups water
3/4 cup sugar
4-1/2 cups sliced rhubarb
1 teaspoon vanilla
1/2 teaspoon cornstarch
1/4 cup cold water

In a 2 quart saucepan dissolve the sugar in the water and bring to a boil. Add rhubarb and reduce heat to low and simmer uncovered for 20 to 30 minutes or until the rhubarb is soft. Remove from heat and stir in vanilla. In a small bowl, mix cornstarch and water to form a paste. Slowly stir this into stewed rhubarb and bring to a boil, stirring constantly. Simmer about 3 to 5 minutes or until thickened. Pour into serving bowl. Top with whipped cream if desired.

Rhubarb Custard Bars

2 cups all-purpose flour
1/4 cup sugar
1 cup cold butter or margarine

Filling:
2 cups sugar
7 tablespoons all-purpose flour
1 cup whipping cream
3 eggs, beaten
5 cups finely chopped fresh or frozen rhubarb thawed and drained

Topping:
2 packages (3 ounces each) cream cheese softened
1/2 cup sugar
1/2 teaspoon vanilla extract
1 cup whipping cream, whipped

In a bowl, combine flour and sugar; cut in the butter until the mixture resembles coarse crumbs. Press into a greased 13 x 9 x 2 baking pan. Bake at 350° for 10 minutes. Meanwhile, for filling, combine sugar and flour in a bowl, whisk in cream and eggs. Stir in the rhubarb and pour over crust. Bake at 350° for 40-45 minutes or until custard is set. Cool. For topping, beat cream cheese, sugar and vanilla until smooth; fold in whipped cream. Spread over top and cover and chill. Cut into bars.
Store in the refrigerator.

Rhubarb Strawberry Apple Crisp

Topping:

3/4 cup rolled oats
1/4 cup whole wheat flour
1/4 cup packed brown sugar
1/2 teaspoon margarine or butter

Filling:

3 cups peeled sliced tart apples
1-1/2 cups sliced strawberries
1-1/2 cups sliced rhubarb
1/2 cup packed brown sugar
2 tablespoons orange juice
1-1/2 tablespoons cornstarch

Preheat oven 375°. In a small bowl combine topping ingredients and set aside. In a large bowl combine the filling ingredients. Coat a shallow 11 x 8 inch baking dish with cooking spray. Pour in filling mixture and spread evenly. Sprinkle with topping and bake until golden brown. Serve warm.

Rhubarb Souffle

1 envelope unflavored gelatin
1/2 cup sugar
1/8 teaspoon salt
4 egg yolks
1/2 cup cold water
1-1/2 cups rhubarb puree
4 egg whites
1/2 cup sugar
1 cup heavy cream, whipped

Tie a foil collar around a 1 quart souffle dish so that the collar extends 2 inches above the rim. Set aside. Combine the gelatin, 1/2 cup sugar and salt in the top of a double boiler. Beat egg yolks and water together until thoroughly blended. Stir into gelatin mixture. Set over boiling water and cook about 5 minutes to cook egg yolks and dissolve gelatin, stirring constantly. Remove from heat and stir in rhubarb puree, cook then chill mixture until slightly thickened, stirring occasionally. Beat egg whites until frothy, add remaining 1/2 cup sugar a little at a time, beating constantly until stiff peaks are formed. Fold whipped cream and meringue together. Fold in chilled gelatin mixture until blended. Turn into prepared dish and chill until firm. When ready to serve carefully remove foil collar.

Rhubarb Tapioca

1 cup pearl tapioca
2 cups cold water
1/4 teaspoon cinnamon
3 cups hot water
2 cups rhubarb
2 cups sugar
1 (14 oz.) can crushed pineapple

Soak the tapioca in the cold water overnight. Add the rest of the ingredients. Over medium low heat cook this mixture stirring constantly until tapioca is clear.

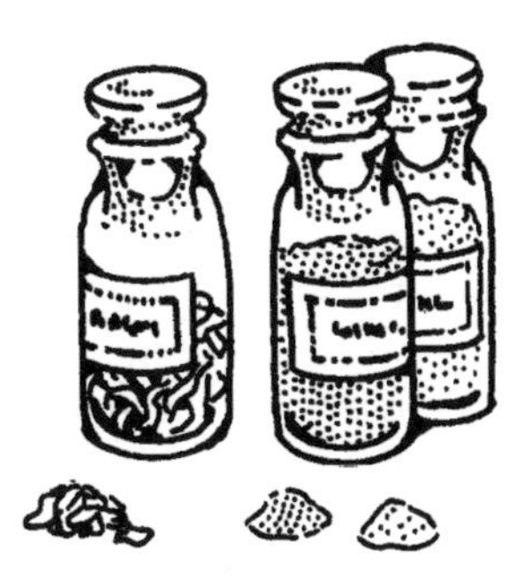

Rhubarb Pudding

1 egg
1 cup sugar
2 cups flour
1 cup milk
2 teaspoons baking powder
1 teaspoon vanilla
4 cups rhubarb, sliced in 1/2 inch pieces
2 cups sugar
2 cups boiling water

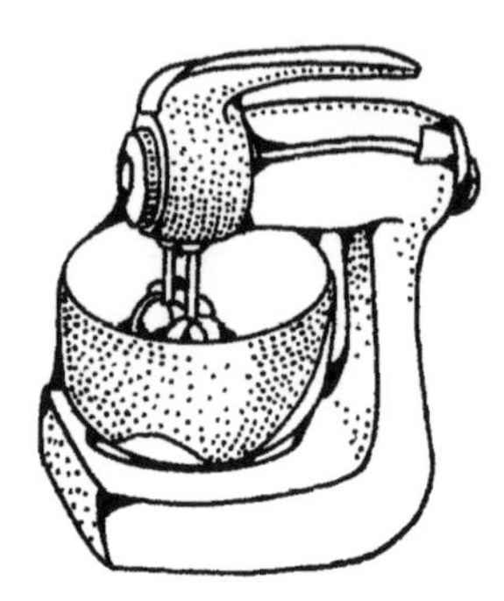

Mix together 1 egg, 1 cup of the sugar, flour, milk, baking powder and vanilla. Place this dough in the bottom on a square cake pan. Then mix together the rhubarb, 2 cups sugar and boiling water. Pour this over the dough and bake at 375° for 40 minutes or until done.

Rhubarb Pudding II

2 cups rhubarb, cut into 1/2 inch pieces
1-1/2 cups sugar
Dash of salt

Thickening:
1/2 cup sugar
1/3 cup flour
2 eggs
1/4 cup cream

In a heavy saucepan over medium heat, cook first 3 ingredients until soft. To make thickening: In a small bowl, using a fork, mix sugar and flour. Stir in the eggs and cream. Add this to the soft rhubarb. Cook over medium heat stirring constantly until thick. Pour into dessert dishes or bowls. Cool or serve hot. Whipped topping if desired.

INDEX

Human hands build a house;
the heart builds a home.

Favorite Recipes

Favorite Recipes

Notes

Notes

Look for Joan Bestwick's *Life's Little Zucchini Cookbook*,
also by Avery Color Studios, Inc.

Avery Color Studios, Inc. has a full line of Great Lakes oriented books, cookbooks, shipwreck and lighthouse maps, and lighthouse posters.

For a full color catalog call:

1-800-722-9925

Avery Color Studios, Inc. products are available at gift shops and bookstores throughout the Great Lakes region.